# Laboratory Manual

# Essentials

## of Biology

*Sylvia S. Mader*

LABORATORY MANUAL TO ACCOMPANY ESSENTIALS OF BIOLOGY

Published by McGraw-Hill, a business unit of The McGraw-Hill Companies, Inc., 1221 Avenue of the Americas, New York, NY 10020. Copyright © 2012 by The McGraw-Hill Companies, Inc. All rights reserved. Previous printings 2010 and 2007. No part of this publication may be reproduced or distributed in any form or by any means, or stored in a database or retrieval system, without the prior written consent of The McGraw-Hill Companies, Inc., including, but not limited to, in any network or other electronic storage or transmission, or broadcast for distance learning.

Some ancillaries, including electronic and print components, may not be available to customers outside the United States.

This book is printed on acid-free paper.

1 2 3 4 5 6 7 8 9 0 QDB/QDB 1 0 9 8 7 6 5 4 3 2 1

ISBN 978–0–07–740215–0
MHID 0–07–740215–4

Vice President, Editor-in-Chief: *Marty Lange*
Vice President, EDP: *Kimberly Meriwether David*
Senior Director of Development: *Kristine Tibbetts*
Publisher: *Janice Roerig-Blong*
Executive Editor: *Michael S. Hackett*
Senior Developmental Editor: *Rose M. Koos*
Senior Marketing Manager: *Tamara Maury*
Project Coordinator: *Mary Jane Lampe*
Senior Buyer: *Sandy Ludovissy*
Senior Designer: *Laurie B. Janssen*
Cover Image: *© Minden Pictures / Masterfile*
Senior Photo Research Coordinator: *Lori Hancock*
Photo Research: *Evelyn Jo Johnson*
Compositor: *Electronic Publishing Services Inc., NYC*
Typeface: *11/13 Slimbach*
Printer: *Quad/Graphics*

All credits appearing on page or at the end of the book are considered to be an extension of the copyright page.

Some of the laboratory experiments included in this text may be hazardous if materials are handled improperly or if procedures are conducted incorrectly. Safety precautions are necessary when you are working with chemicals, glass test tubes, hot water baths, sharp instruments, and the like, or for any procedures that generally require caution. Your school may have set regulations regarding safety procedures that your instructor will explain to you. Should you have any problems with materials or procedures, please ask your instructor for help.

www.mhhe.com

# Contents

# Preface

## To the Instructor

The laboratory exercises in this manual are coordinated with *Essentials of Biology*, a general biology text by Sylvia S. Mader that covers principles of biology from an evolutionary perspective.

Although each laboratory is referenced to the appropriate chapter in *Essentials of Biology*, this manual may also be used in coordination with other general biology texts. In addition, this laboratory manual can be adapted to a variety of course orientations and designs. There are a sufficient number of laboratories and activities within each lab to tailor the laboratory experience as desired. Then, too, many activities may be performed as demonstrations rather than as student activities, thereby shortening the time required to cover a particular concept.

### The Exercises

All exercises have been tested for student interest, preparation time, estimated time of completion, and feasibility. The following features are particularly appreciated by adopters:

**Integrated opening:** Each laboratory begins with a list of learning objectives organized according to the major sections of the laboratory. The major sections of the laboratory are numbered on the opening page and in the laboratory text material. This organization will help students better understand the goals of each laboratory session.

**Self-contained content:** Each laboratory contains all the background information necessary to understand the concepts being studied and to answer the questions asked. This feature will reduce student frustration and increase learning.

**Scientific process:** All laboratories stress the scientific process, and many opportunities are given for students to gain an appreciation of the scientific method. The first laboratory of this edition explicitly explains the steps of the scientific method and gives students an opportunity to use them. I particularly recommend this laboratory because it uses the pillbug, a living subject.

**Student activities:** A color bar is used to designate each student activity. Some student activities are Observations and some are Experimental Procedures. An icon appears whenever a procedure requires time before results can be viewed. Sequentially numbered steps guide students as they perform each activity.

**Live materials:** Although students work with living material during some part of almost all laboratories, the exercises are designed to be completed within one laboratory session. This facilitates the use of the manual in multiple-section courses.

**Laboratory safety:** Laboratory safety is of prime importance, and the listing on page vii will assist instructors in making the laboratory experience a safe one.

### Customized Editions

The 21 laboratories in this manual are available as individual "lab separates," so instructors can custom-tailor the manual to their particular course needs.

### Laboratory Resource Guide

The *Laboratory Resource Guide,* an essential aid for instructors and laboratory assistants, is available online free to adopters of the *Laboratory Manual*. The answers to the Laboratory Review questions are in the *Resource Guide.*

# To the Student

Special care has been taken in preparing the *Laboratory Manual for Essentials of Biology* to enable you to *enjoy* the laboratory experience as you *learn* from it. The instructions and discussions are written clearly so that you can understand the material while working through it. Student aids are designed to help you focus on important aspects of each exercise.

## Student Learning Aids

Student learning aids are carefully integrated throughout this manual. The Learning Objectives set the goals of each laboratory session and help you review the material for a laboratory practical or any other kind of exam. In this edition, the major sections of each laboratory are numbered, and the learning objectives are grouped according to these topics. This system allows you to study the chapter in terms of the objectives presented.

The Introduction reviews much of the necessary background information required for comprehending upcoming experiments. Color bars bring attention to exercises that require your active participation by highlighting Observations and Experimental Procedures, and an icon indicates a timed experiment. Throughout, space is provided for recording answers to questions and the results of investigations and experiments. Each laboratory ends with a set of review questions covering the day's work.

Appendix A at the end of the book provides useful information on the metric system. Practical examination answer sheets are also provided.

## Laboratory Preparation

*Read* each exercise before coming to the laboratory. *Study* the introductory material and the procedures. To obtain a better understanding, read the corresponding chapter in your text. If your text is *Essentials of Biology* by Sylvia S. Mader, see the "text chapter reference" column in the table of contents at the beginning of this *Laboratory Manual.*

## Color Bar and Icon

Throughout each laboratory, a color bar designates either an Observation or an Experimental Procedure.

- **Observation:** An activity in which models, slides, and preserved or live organisms are observed to achieve a learning objective.
- **Experimental Procedure:** An activity in which a series of steps uses laboratory equipment to gather data and come to a conclusion.
- **Time:** An icon is used to designate when time is needed for an Experimental Procedure. Allow the designated amount of time for this activity. Start these activities at the beginning of the laboratory, proceed to other activities, and return to these when the designated time is up.

## Laboratory Review

Each laboratory ends with a number of thought questions that will help you determine whether you have accomplished the objectives for the laboratory.

## Student Feedback

If you have any suggestions for how this laboratory manual could be improved, you may send your comments to:

The McGraw-Hill Companies
Product Development—General Biology
501 Bell Street
Dubuque, Iowa 52001

or use the Feedback form on the *Essentials of Biology* website at www.mhhe.com/maderessentials3.

# Acknowledgments

Most of the exercises in this laboratory manual are revised from other laboratory manuals by Dr. Sylvia Mader. As such, they have been reviewed by instructors who use those manuals. However, we gratefully acknowledge the following instructors who suggested the exercises that should be included in this manual.

**Kelly S. Cartwright,** *College of Lake County*

**Aimée Lee,** *The University of Southern Mississippi*

**Sarah Meiers,** *Western Illinois University*

**Kirsten Raines,** *San Jacinto College*

**Jennifer Warner,** *University of North Carolina—Charlotte*

# Laboratory Safety

The following is a list of practices required for safety purposes in the biology laboratory and in outdoor activities. Following rules of lab safety and using common sense throughout the course will enhance your learning experience by increasing your confidence in your ability to safely use chemicals and equipment. Pay particular attention to oral and written safety instructions given by the instructor. If you do not understand a procedure, ask the instructor, rather than a fellow student, for clarification. Be aware of your school's policy regarding accident liability and any medical care needed as a result of a laboratory or outdoor accident.

**The following rules of laboratory safety should become a habit:**

1. Wear safety glasses or goggles during exercises in which glassware and solutions are heated or when dangerous fumes may be present, creating possible hazards to eyes or contact lenses.
2. Assume that all reagents are poisonous and act accordingly. Read the labels on chemical bottles for safety precautions and know the nature of the chemical you are using. If chemicals come into contact with skin, wash immediately with water.
3. **DO NOT**
   a. ingest any reagents.
   b. eat, drink, or smoke in the laboratory. Toxic material may be present, and some chemicals are flammable.
   c. carry reagent bottles around the room.
   d. pipette anything by mouth.
   e. put chemicals in the sink or trash unless instructed to do so.
   f. pour chemicals back into containers unless instructed to do so.
   g. operate any equipment until you are instructed in its use.
4. **DO**
   a. note the location of emergency equipment such as a first aid kit, eyewash bottle, fire extinguisher, switch for ceiling showers, fire blanket(s), sand bucket, and telephone (911).
   b. be familiar with the experiments you will be doing before coming to the laboratory. This will increase your understanding, enjoyment, and safety during exercises. Confusion is dangerous. Completely follow the procedure set forth by the instructor.
   c. keep your work area neat, clean, and organized. Before beginning, remove everything from your work area except the lab manual, pen, and equipment used for the experiment. Wash hands and desk area, including desk top and edge, before and after each experiment. Use clean glassware at the beginning of each exercise, and wash glassware at the end of each exercise or before leaving the laboratory.
   d. wear clothing that, if damaged, would not be a serious loss, or use aprons or laboratory coats, since chemicals may damage fabrics.
   e. wear shoes as protection against broken glass or spillage that may not have been adequately cleaned up.
   f. handle hot glassware with a test tube clamp or tongs. Use caution when using heat, especially when heating chemicals. Do not leave a flame unattended; do not light a Bunsen burner near a gas tank or cylinder; do not move a lit Bunsen burner; do keep long hair and loose clothing well away from the flame; do make certain gas jets are off when the Bunsen burner is not in use. Use proper ventilation and hoods when instructed.
   g. read chemical bottle labels; be aware of the hazards of all chemicals used. Know the safety precautions for each.
   h. stopper all reagent bottles when not in use. Immediately wash reagents off yourself and your clothing if they spill on you, and immediately inform the instructor. If you accidentally get any reagent in your mouth, rinse the mouth thoroughly, and immediately inform your instructor.
   i. use extra care and wear disposable gloves when working with glass tubing and when using dissection equipment (scalpels, knives, or razor blades), whether cutting or assisting.
   j. administer first aid immediately to clean, sterilize, and cover any scrapes, cuts, and burns where the skin is broken and/or where there may be bleeding. Wear bandages over open skin wounds.
   k. report all accidents to the instructor immediately, and ask your instructor for assistance in cleaning up broken glassware and spills.
   l. report to the instructor any condition that appears unsafe or hazardous.
   m. use caution during any outdoor activities. Watch for snakes, poisonous insects or spiders, stinging insects, poison oak, poison ivy, and so on. Be careful near water.

**I understand the safety rules as presented above. I agree to follow them and all other instructions given by the instructor.**

Name: _____     Date: _____

Laboratory Class and Time: _____

# 1

# Scientific Method

## Introduction

This laboratory will provide you with an opportunity to use the scientific method in the same manner as scientists. Scientists often begin by making observations about the subject of interest. Today our subject is the pillbug, *Armadillidium vulgare,* a type of crustacean that lives on land (Fig. 1.1).

Pillbugs have overlapping "armored" plates that make them look like little armadillos. A pillbug can roll up into such a tight ball that its legs and head are no longer visible, earning it the nickname "roly-poly." They are commonly found in damp leaf litter, under rocks, and in basements or crawl spaces under houses. Pillbugs breathe by gills located on the underside of their bodies. The gills must be kept slightly moist, and that is why they are usually found in damp places.

In winter, pillbugs are inactive, but when spring arrives they become active and mate. Females have a pouch on the underside of their body, where they can carry from 7 to 200 eggs. The eggs hatch several weeks after mating, and the young look like miniature adults. The young stay in the pouch another six weeks, and then they leave and begin to feed. They eat decaying plants and animals and some living plants. They can live up to three years.

**Figure 1.1  Pillbugs (roly-poly bugs).**
The pillbug, *A. vulgare,* lives on land. They are wingless, and coloration varies from brown to slate grey. They breathe by gills, and they have seven pairs of jointed legs for locomotion on land. Pillbugs live in damp places, mostly under rocks, wood, and decaying leaves. They feed at night.

Pillbugs molt (shed the exoskeleton) four or five times. They have three body parts: head, thorax, and abdomen. Among crustaceans, pillbugs are classified as isopods because they are dorsoventrally flattened, lack a carapace, have compound eyes, and have two pairs of antennae.

Isopods are the only crustaceans that include forms adapted to living their entire life on land, although moisture is required. Currently, it is believed that they do not transmit diseases, nor do they bite or sting. Because they eat dead organic matter, such as leaves, they are easy to find and keep in a moist terrarium with leaf litter, rocks, and wood chips. You are encouraged to collect some for your experiment. Since they live in the same locations as snakes, be careful when collecting them.

First, become acquainted with your subject and how it normally moves. Then, you will use your knowledge of the pillbug to hypothesize whether it will be attracted to, repelled by, or indifferent to various substances of your choice. After you have tested your hypotheses, you will conclude whether they are supported. Finally, your conclusions may lead to other hypotheses, and if time permits, you may go ahead and test those also.

## 1.1 Using the Scientific Method

Scientists use the scientific method (Fig. 1.2) to come to a conclusion about the natural world. When a scientist begins a study, he or she uses preliminary observations and previous data to formulate a hypothesis. A **hypothesis** is a tentative explanation of observed phenomena.

After a hypothesis is formulated, it must be tested by doing new experiments and/or making new observations. Experiments are done and observations made in such a way that others can repeat them. Only repeatable observations and experiments are accepted as valid contributions to the field of science.

**Data** are any factual information that can be observed either independently of, or as a result of, experimentation. On the basis of the data, a scientist comes to a **conclusion**—whether the observations support the hypothesis or prove it false. Scientists are always aware that further observations and experiments could lead to a change in prior conclusions. Therefore, it is never said that the data prove a hypothesis true, but we could say that the data support the hypothesis. The arrow in Figure 1.2 indicates that research often enters a cycle of hypothesis—experiments and observations—conclusion—hypothesis. Explain this cycle. _____

_____

After many years of testing and study, the scientific community may develop a **scientific theory,** a concept that ties together many varied conclusions into a generalized statement. For example, after testing the cause of many individual diseases, the germ theory of disease was formulated. It states that infectious diseases are caused by pathogens (e.g., bacteria and viruses) that can be passed from one person to another. How is a scientific theory different from a conclusion?

_____

_____

**Figure 1.2** **Flow diagram for the scientific method.**
On the basis of observations, a scientist formulates a hypothesis. The hypothesis is tested by further experiments and/or observations. The scientist then concludes whether the data support or do not support the hypothesis. If the data does not support the hypothesis, the hypothesis is rejected and the scientist starts again. If the hypothesis is supported, scientists continue to work and eventually may formulate a scientific theory.

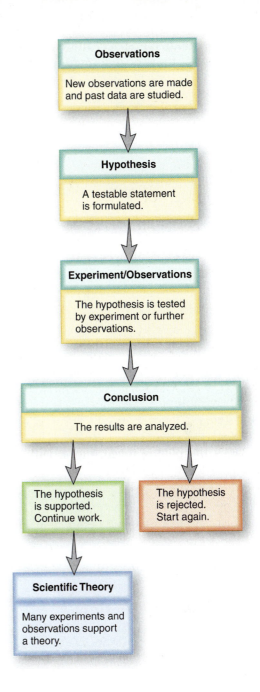

**Observations**
New observations are made and past data are studied.

**Hypothesis**
A testable statement is formulated.

**Experiment/Observations**
The hypothesis is tested by experiment or further observations.

**Conclusion**
The results are analyzed.

The hypothesis is supported. Continue work.

The hypothesis is rejected. Start again.

**Scientific Theory**
Many experiments and observations support a theory.

# 1.2 Observing the Pillbug

Wash your hands before and after handling pillbugs. Please handle them carefully so they are not crushed. When touched, they roll up into a ball or "pill" shape as a defense mechanism. They will soon recover if left alone.

*Observation: Pillbug's External Anatomy*

1. Obtain a pillbug that has been numbered with white correction fluid or tape tags. First examine the shell and body with the unaided eye and then with a magnifying lens or dissecting microscope. Put the pillbug in a small glass or plastic dish to keep it contained.
2. Examine the shell shape, color, and texture. Note the number of legs and antennae and whether there are any posterior appendages, such as uropods (paired appendages at end of abdomen) or brood pouches. (Females have leaflike growths at the base of some legs where developing eggs and embryos are held in pouches.) Locate the eyes. Count the number of overlapping plates.

3. In the following space, draw a large outline of your pillbug (at least 10–12 cm across). Label the head, thorax, abdomen, antennae, eyes, uropods, and one of the seven pairs of legs.

4. Draw the pillbug rolled into a ball.

## Observation: Pillbug's Motion

1.  Watch a pillbug's underside as the pillbug moves up a transparent surface, such as the side of a graduated cylinder or beaker. Describe the action of the feet and any other motion you see.

    _____

2.  As you watch the pillbug, identify behaviors that might

    a. protect it from predators _____

    b. help it acquire food _____

    c. protect it from the elements _____

    d. allow interaction with the environment _____

3.  Allow a pillbug to crawl on your hand. Describe how it feels and how it acts.

    _____

4.  Place the pillbug on a graduated cylinder. Experiment with the angle of the cylinder and the position of the pillbug to determine the pillbug's preferred direction of motion. For example, place the cylinder on end, and position the pillbug so that it can move up or down. Try other arrangements also. Repeat this procedure with three other pillbugs. In Table 1.1, record the preferred direction of motion and other observations for each pillbug.

| Table 1.1 | Preferred Direction of Motion | |
| --- | --- | --- |
| Pillbug | Direction Moved | Comments |
| 1 | | |
| 2 | | |
| 3 | | |
| 4 | | |

5.  Measure the speed of the pillbug. Use what you learned about each pillbug's preferred direction of motion (see Table 1.1) to get maximum cooperation from each of the four pillbugs you worked with in step 4. Place each pillbug on a metric ruler, and use a stopwatch to measure the time it takes for the pillbug to move a certain number of centimeters. In Table 1.2, record your results for each pillbug. Calculate each pillbug's average speed in millimeters (mm) per second here, and record your data in Table 1.2.

| Table 1.2 | Pillbug Speed | | |
| --- | --- | --- | --- |
| Pillbug | Millimeters Traveled | Time (sec) | Average Speed (mm/sec) |
| 1 | | | |
| 2 | | | |
| 3 | | | |
| 4 | | | |

# 1.3 Formulating Hypotheses

Hypotheses are often stated as "if-then" statements. For example, *if* the pillbug is exposed to _____, *then* it will be _____. You will be testing whether pillbugs are attracted to, repelled by, or unresponsive to particular substances. Pillbugs move away from a substance when they are repelled by it, and they move toward and eat a substance they are attracted to. If a pillbug simply rolls into a ball, nothing can be concluded, and you may wish to choose another pillbug or wait a minute or two to check for further response.

1. Choose

   a. two or three powders, such as flour, cornstarch, coffee creamer, baking soda, fine sand.
   b. two or three liquids, such as milk, orange juice, ketchup, applesauce, a carbonated beverage, water.

2. In Table 1.3 write a hypothesis about how you expect the pillbug to respond, and offer an explanation for your reasoning.

| Table 1.3 | Hypotheses About Pillbug's Reaction to Common Powders and Liquids | | |
|---|---|---|---|
| Substance Tested | Hypothesis About How Pillbug Will Respond to Substance | | Reasoning for Hypothesis |
| 1 | | | |
| 2 | | | |
| 3 | | | |
| 4 | | | |
| 5 | | | |
| 6 | | | |

# 1.4 Performing an Experiment

Design an experiment to test the pillbug's reaction to the chosen substances. The pillbug must be treated humanely. No substance must be put directly on the pillbug, nor can the pillbug be placed directly onto the substance. Since pillbugs tend to walk around the edge of a petri dish, you could put the wet or dry substance around the edge of the dish. Or for wet substances, you could put liquid-soaked cotton in the pillbug's path.

A good experimental design contains a **control.** A control group goes through all the steps of an experiment but lacks, or is not exposed to, the factor being tested. If you are testing the pillbug's reaction to a liquid, water can be the control substance substituted for the test liquid. If you are testing the pillbug's reaction to a powder, substitute fine sand for the test powder.

### Experimental Procedure: Pillbug's Reaction to Common Substances

1. What substances are you testing? Include in your list any controls, and complete the first column in Table 1.4.
2. Rinse your pillbug between procedures by spritzing with distilled water from a spray bottle. Then put it on a paper towel to dry off.
3. Test the pillbug's reaction using the previously described method.

4. Watch the pillbug's reaction to each substance, and record it in Table 1.4.
5. Do your results support your hypotheses? Answer *yes* or *no* in the last column.

| Table 1.4 | Pillbug's Reaction to Common Substances | |
|---|---|---|
| **Substance Tested** | **Pillbug's Reaction** | **Hypothesis Supported?** |
| 1 | | |
| 2 | | |
| 3 | | |
| 4 | | |
| 5 | | |
| 6 | | |

6. Compare your results with those of other students who tested the same substances. In Table 1.5, list any substances which attracted pillbugs.

| Table 1.5 | | | | Class Results | | | | | |
|---|---|---|---|---|---|---|---|---|---|
| **Group No.** | **Exp. 1 Direction** | **Exp. 2 Speed (mm/sec)** | **Exp. 3 Dry Control** | *Dry Substances* | | | **Exp. 4 Wet Control** | *Wet Substances* | |
| 1 | | | | | | | | | |
| 2 | | | | | | | | | |
| 3 | | | | | | | | | |
| 4 | | | | | | | | | |
| 5 | | | | | | | | | |
| 6 | | | | | | | | | |
| 7 | | | | | | | | | |
| 8 | | | | | | | | | |

*Continuing the Experiment*
7. Study your results and those of other students, and decide what factors may have caused the pillbug to be attracted to or repelled by a substance. _____

_____

On the basis of your decision, what is your new hypothesis? _____

_____

**8.** Test your hypothesis, and describe your results here. If possible, make up a table to display your results.

**9.** Based on your new data, what is your conclusion?

_____

## Laboratory Review 1

_____    **1.** Which is more comprehensive, a conclusion or a theory?

_____    **2.** What is a tentative explanation of observed phenomena?

_____    **3.** What do you call the information scientists collect when doing experiments and making observations?

_____    **4.** What step in the scientific method follows experiments and observations?

_____    **5.** What do you call a sample that goes through all the steps of an experiment and does not contain the factor being tested?

_____    **6.** Can data prove a hypothesis true? (Yes or No)

Indicate whether statements 7 and 8 are hypotheses, conclusions, or theories.

_____    **7.** The data show that vaccines protect people from disease.

_____    **8.** All living things are made of cells.

_____    **9.** What part of a pillbug is for protection?

_____    **10.** What does a pillbug do to protect itself?

_____    **11.** How many body divisions does a pillbug have?

_____    **12.** If a pillbug travels 1 cm in two minutes, what is its rate of speed?

_____    **13.** What can be concluded if a pillbug curls into a ball?

_____    **14.** Pillbugs that back away from a substance are (attracted to/repelled by) the substance.

### Thought Questions

**15.** What is a theory?

**16.** Why is it important to use one substance at a time when testing a pillbug's reaction?

# 2

# Metric Measurement and Microscopy

## Learning Objectives

**2.1 The Metric System**
- Use metric units of measurement for length and volume.
- Convert between metric units using an understanding of metric prefixes.

**2.2 Microscopy**
- Describe differences between the compound light microscope and the electron microscope.

**2.3 Binocular Dissecting Microscope (Stereomicroscope)**
- Identify the parts and tell how to focus the binocular dissecting microscope.

**2.4 Use of the Compound Light Microscope**
- Name and give the function of the basic parts of the compound light microscope.
- List, in proper order, the steps for bringing an object into focus with the compound light microscope.
- Describe how a slide of any object provides information on the inversion of the image by the compound light microscope.
- Calculate the total magnification for the various lens systems.

**2.5 Microscopic Observations**
- Name the three kinds of cells studied in this exercise.
- State two differences between onion epidermal cells and human epithelial cells.

## Introduction

This laboratory introduces you to the metric system, which biologists use to indicate the sizes of cells and cell structures. This laboratory also examines the features, functions, and use of the compound light microscope and the binocular dissecting microscope. Transmission and scanning electron microscopes are explained, and micrographs produced using these microscopes appear throughout this lab manual. The binocular dissecting microscope and the scanning electron microscope view the surface and/or the three-dimensional structure of an object. The compound light microscope and the transmission electron microscope can view only extremely thin sections of a specimen. If the cell in Figure 2.1a was sectioned lengthwise for viewing, the interior of the projections at the top of the cell, called cilia, would appear in the micrograph. A lengthwise cut through any type of specimen is called a **longitudinal section (l.s.)**. On the other hand, if the cell was sectioned crosswise below the area of the cilia, you would see other portions of the interior of the subject. A crosswise cut through any type of specimen is called a **cross section (c.s.)**.

**Figure 2.1  Longitudinal and cross sections.**
*a.* Transparent view of a cell. *b.* A longitudinal section would show the cilia at the top of the cell. *c.* A cross section shows only the interior where the cut is made.

a. Cell

b. Longitudinal section

c. Cross section

# 2.1 The Metric System

The **metric system** is the standard system of measurement in the sciences, including biology, chemistry, and physics. It has tremendous advantages because all conversions, whether for volume, mass (weight), or length, are conveniently represented in units of ten. Refer to text for an in-depth look at the units.

## Length

Metric units of length measurement studied in this laboratory include the **meter** (m), **centimeter** (cm), **millimeter** (mm), and **micrometer** (μm) (Table 2.1). The prefixes centi- ($10^{-2}$), milli- ($10^{-3}$), micro- ($10^{-6}$), and nano- ($10^{-9}$) are used with length, weight, and volume.

| Table 2.1 | Metric Units of Length Measurement | | |
|---|---|---|---|
| **Unit** | **Meters** | **Centimeters** | **Millimeters** |
| Meter (m) | 1 m | 100 cm | 1,000 mm |
| Centimeter (cm) | 0.01 ($10^{-2}$) m | 1 cm | 10 mm |
| Millimeter (mm) | 0.001 ($10^{-3}$) m | 0.1 cm | 1.0 mm |
| Micrometer (μm) | 0.000001 ($10^{-6}$) m | 0.0001 ($10^{-4}$) cm | 0.001 ($10^{-3}$) mm |
| Nanometer (nm) | 0.000000001 ($10^{-9}$) m | 0.0000001 ($10^{-7}$) cm | 0.000001 ($10^{-6}$) mm |

### Experimental Procedure: Length

1. Obtain a small ruler marked in centimeters and milli-
meters. How many centimeters are represented? _____
One centimeter equals how many millimeters? _____
To express the size of small objects, such as cell contents, biologists use even smaller units of the metric system than those on the ruler. These units are the micrometer (μm) and the nanometer (nm).
According to Table 2.1, 1 μm = _____ mm, and 1 nm = _____ mm.
Therefore, 1 mm = _____ μm = _____ nm.

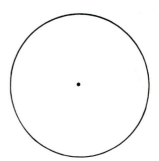

2. Measure the diameter of the circle shown to the nearest millimeter. This circle is

   _____ mm = _____ μm = _____ nm.

## Volume

Two metric units of volume are the **liter** (l) and the **milliliter** (ml). One liter = 1,000 ml.

### Experimental Procedure: Volume

1. Volume measurements can be related to those of length. For example, use a millimeter ruler to measure a wooden block to get its length, width, and depth.

   length = _____ cm; width = _____ cm; depth = _____ cm

The volume, or space, occupied by the wooden block can be expressed in cubic centimeters (cc or cm³) by multiplying: length × width × depth = _____ cm³. For purposes of this Experimental Procedure, 1 cubic centimeter equals 1 milliliter; therefore, the wooden block has a volume of _____ ml.

2. In the biology laboratory, liquid volume is usually measured directly in liters or milliliters with appropriate measuring devices. For example, use a 50 ml graduated cylinder to add 20 ml of water to a test tube. First, fill the graduated cylinder to the 20 ml mark. To do this properly, you have to make sure that the lowest margin of the water level, or the **meniscus** (Fig. 2.2), is at the 20 ml mark. Place your eye directly parallel to the level of the meniscus, and add water until the meniscus is at the 20 ml mark. (Having a dropper bottle filled with water on hand can help you do this.) A large, blank, white index card held behind the cylinder can also help you see the scale more clearly. Now, pour the 20 ml of water into the test tube.

3. Hypothesize how you could find the total volume of the test tube. _____

_____

Now perform the operation you suggested. What is the test tube's total volume? _____

4. Fill a 50 ml graduated cylinder with water to about the 20 ml mark. How could you use this setup to calculate the volume of the wooden block? _____

_____

Now perform the operation you suggested. With this method, the wooden block has a volume of _____ ml. Do your two values agree? _____ Explain. _____

_____

5. How could you determine how many drops from the pipette of the dropper bottle equal 1 ml?

_____

Now perform the operation you suggested. How many drops from the pipette of the dropper bottle equal 1 ml? _____ Some pipettes are graduated and can be filled to a certain level as a way to measure volume directly. Your instructor will demonstrate this. Are pipettes customarily used to measure large or small volumes? _____

**Figure 2.2 Meniscus.**
The proper way to view the meniscus.

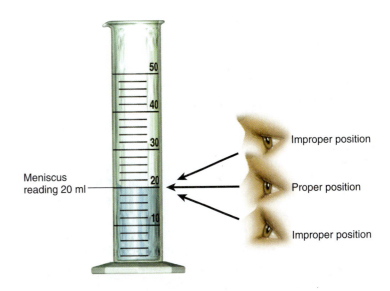

Improper position

Proper position

Improper position

Meniscus
reading 20 ml

Metric Measurement and Microscopy   Laboratory 2   **11**

## 2.2 Microscopy

Because biological objects can be very small, we often use a microscope to view them. Many kinds of instruments, ranging from the hand lens to the electron microscope, are effective magnifying devices. A short description of two kinds of light microscopes and two kinds of electron microscopes follows.

### Light Microscopes

Light microscopes use visible light rays that are magnified and focused by means of lenses. The **binocular dissecting microscope** (stereomicroscope) is designed to study entire objects in three dimensions at low magnification. The **compound light microscope** is used for examining small or thinly sliced sections of objects under higher magnification than that of the binocular dissecting microscope. The term **compound** refers to the use of two sets of lenses: the ocular lenses located near the eyes and the objective lenses located near the object. Illumination is from below, and visible light passes through clear portions but does not pass through opaque portions. To improve contrast, the microscopist uses stains or dyes that bind to cellular structures and absorb light. Figure 2.3*a* is a **photomicrograph,** a photograph of an image produced by a compound light microscope.

### Figure 2.3 Comparative micrographs.

Micrographs of a lymphocyte, a type of white blood cell. The photomicrograph in (*a*) shows less detail than the transmission electron micrograph (TEM) in (*b*). The scanning electron micrograph (SEM) in (*c*) shows the cell surface in three dimensions.

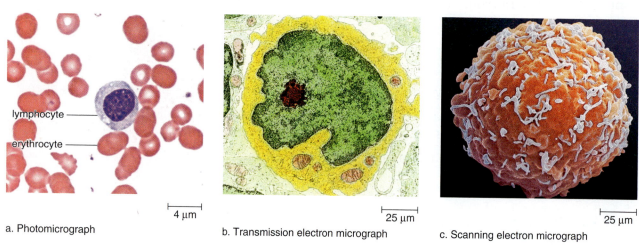

lymphocyte

erythrocyte

|—————| 4 µm

|—————| 25 µm

|—————| 25 µm

a. Photomicrograph      b. Transmission electron micrograph      c. Scanning electron micrograph

### Rules for Microscope Use

**Observe the following rules for using a microscope:**

1. The lowest power objective (scanning or low) should be in position both at the beginning and at the end of microscope use.
2. Use only lens paper for cleaning lenses.
3. Do not tilt the microscope as the eyepieces could fall out or wet mounts could be ruined.
4. Keep the stage clean and dry to prevent rust and corrosion.
5. Do not remove parts of the microscope.
6. Keep the microscope dust-free by covering it after use.
7. Report any malfunctions.

## Electron Microscopes

Electron microscopes use a beam of electrons magnified and focused on a photographic plate by means of electromagnets. The **transmission electron microscope** is analogous to the compound light microscope. The object is ultra-thinly sliced and treated with heavy metal salts to provide contrast. Figure 2.3*b* is a micrograph produced by this type of microscope. The **scanning electron microscope** is analogous to the dissecting light microscope. It gives an image of the surface and dimensions of an object, as is apparent from the scanning electron micrograph in Figure 2.3*c*.

The micrographs in Figure 2.3 demonstrate that much smaller objects can be viewed with electron microscopes than with compound light microscopes. The difference between these two types of microscopes, however, is not simply a matter of magnification; it is also the electron microscope's ability to show detail. The electron microscope has greater resolving power. **Resolution** is the minimum distance between two objects at which they can still be seen, or resolved, as two separate objects. The use of high energy electrons rather than visible light gives electron microscopes a much greater resolving power since two objects that are much closer together can still be distinguished as separate points. Table 2.2 lists several other differences between the compound light microscope and the transmission electron microscope.

| Table 2.2 | Comparison of the Compound Light Microscope and the Transmission Electron Microscope |
|---|---|
| **Compound Light Microscope** | **Transmission Electron Microscope** |
| 1. Glass lenses | 1. Electromagnetic lenses |
| 2. Illumination by visible light | 2. Illumination due to a beam of electrons |
| 3. Resolution $\cong$ 200 nm | 3. Resolution $\cong$ 0.1 nm |
| 4. Magnifies to 1,500× | 4. Magnifies to 1,000,000× |
| 5. Costs up to tens of thousands of dollars | 5. Costs up to hundreds of thousands of dollars |

### Conclusions

- Which two types of microscopes view the surface of an object? _____
- Which two types of microscopes view objects that have been sliced and treated to improve contrast? _____
- Of the microscopes just mentioned, which one resolves the greater amount of detail?

_____

## 2.3 Binocular Dissecting Microscope (Stereomicroscope)

The **binocular dissecting microscope** allows you to view objects in three dimensions at low magnifications. It is used to study entire small organisms, any object requiring lower magnification, and opaque objects that can be viewed only by reflected light. It is also called a stereomicroscope because it produces a three-dimensional image.

### Identifying the Parts

After your instructor has explained how to carry a microscope, obtain a binocular dissecting microscope and a separate illuminator, if necessary, from the storage area. Place it securely on the table. Plug in the power cord and turn on the illuminator. There is a wide variety of binocular dissecting microscope styles, and your instructor will discuss the specific style(s) available to you. Regardless of style, the features on the following page should be present:

**Figure 2.4  Binocular dissecting microscope (stereomicroscope).**
Label this microscope with the help of the text material.

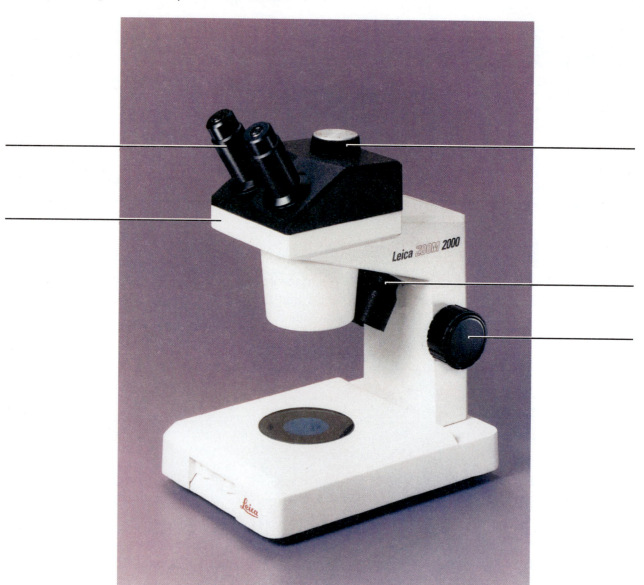

1. **Binocular head:** Holds two eyepiece lenses that move to accommodate for the various distances between different individual's eyes.

2. **Eyepiece lenses:** The two lenses located on the binocular head. What is the magnification of your eyepieces? _____ Some models have one **independent focusing eyepiece** with a knurled knob to allow independent adjustment of each eye. The nonadjustable eyepiece is called the **fixed eyepiece.**

3. **Focusing knob:** A large black or gray knob located on the arm; used for changing the focus of both eyepieces together.

4. **Magnification changing knob:** A knob, often built into the binocular head, used to change magnification in both eyepieces simultaneously. This may be a **zoom** mechanism or a **rotating lens** mechanism of different powers that clicks into place.

5. **Illuminator:** Used to illuminate an object from above; may be built into the microscope or separate.

*Locate each of these parts on your binocular dissecting microscope, and label them on Figure 2.4.*

## Focusing the Binocular Dissecting Microscope

1. Place a plastomount that contains small organisms in the center of the stage.
2. Adjust the distance between the eyepieces on the binocular head so that they comfortably fit the distance between your eyes. You should be able to see the object with both eyes as one three-dimensional image.
3. Use the focusing knob to bring the object into focus.
4. Does your microscope have an independent focusing eyepiece? _____ If so, use the focusing knob to bring the image in the fixed eyepiece into focus, while keeping the eye at the independent focusing eyepiece closed. Then adjust the independent focusing eyepiece so that the image is clear, while keeping the other eye closed. Is the image inverted? _____
5. Turn the magnification changing knob, and determine the kind of mechanism on your microscope. A zoom mechanism allows continuous viewing while changing the magnification. A rotating lens mechanism blocks the view of the object as the new lenses are rotated. Be sure to click each lens firmly into place. If you do not, the field will be only partially visible. What kind of mechanism is on your microscope? _____
6. Set the magnification changing knob on the lowest magnification. Sketch the object in the following circle as though this represents your entire field of view:

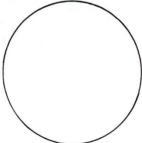

7. Rotate the magnification changing knob to the highest magnification. Draw another circle within the one provided to indicate the reduction of the field of view.
8. Experiment with various objects at various magnifications until you are comfortable with using the binocular dissecting microscope.
9. When you are finished, return your binocular dissecting microscope and illuminator to their correct storage areas.

# 2.4 Use of the Compound Light Microscope

As mentioned, the name **compound light microscope** indicates that it uses two sets of lenses and light to view an object. The two sets of lenses are the ocular lenses located near the eyes and the objective lenses located near the object. Illumination is from below, and the light passes through clear portions but does not pass through opaque portions. This microscope is used to examine smaller objects or more thinly sliced sections of objects under higher magnification than would be possible with the binocular dissecting microscope.

## Identifying the Parts

Obtain a compound light microscope from the storage area and place it securely on the table. *Identify the following parts on your microscope, and label them in Figure 2.5.*

**Figure 2.5** Compound light microscope.
Compound light microscope with binocular head and mechanical stage. Label this microscope with the help of the text material.

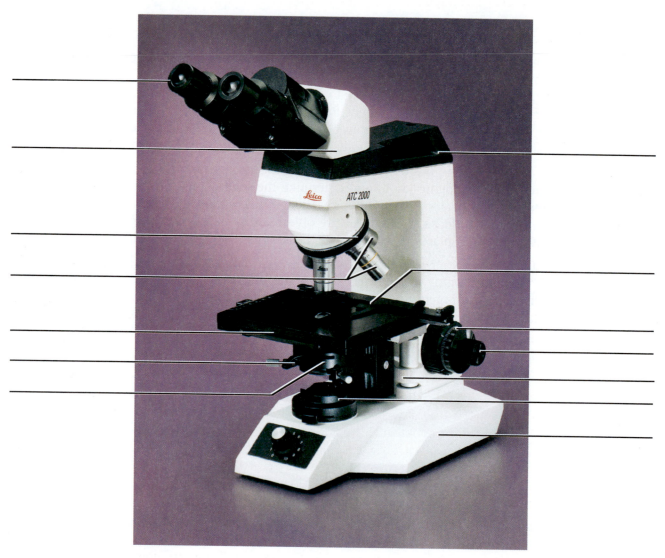

1. **Eyepieces (ocular lenses):** What is the magnifying power of the ocular lenses on your microscope? _____

2. **Body tube:** Holds nosepiece at one end and eyepiece at the other end; conducts light rays.

3. **Arm:** Supports upper parts and provides carrying handle.

4. **Nosepiece:** Revolving device that holds objectives.

5. **Objectives** (objective lenses):

   a. **Scanning power objective:** This is the shortest of the objective lenses and is used to scan the whole slide. The magnification is stamped on the housing of the lens. It is a number followed by an ×. What is the magnifying power of the scanning lens on your microscope? _____

   b. **Low-power objective:** This lens is longer than the scanning lens and is used to view objects in greater detail. What is the magnifying power of the low-power objective lens on your microscope? _____

   c. **High-power objective:** If your microscope has three objective lenses, this lens will be the longest. It is used to view an object in even greater detail. What is the magnifying power of the high-power objective lens on your microscope? _____

   d. **Oil immersion objective** (on microscopes with four objective lenses): Holds a 95× (to 100×) lens and is used in conjunction with immersion oil to view objects with the greatest magnification. Does your microscope have an oil immersion objective? _____ If this lens is available, your instructor will discuss its use when the lens is needed.

6. **Coarse-adjustment knob:** Knob used to bring object into approximate focus; used only with scanning and low-power objectives.

7. **Fine-adjustment knob:** Knob used to bring object into final focus.

8. **Condensor:** Lens system below the stage used to focus the beam of light on the object being viewed.

9. **Diaphragm** or **diaphragm control lever:** Controls amount of illumination used to view the object.

10. **Light source:** An attached lamp that directs a beam of light up through the object.

11. **Base:** The flat surface of the microscope that rests on the table.

12. **Stage:** Holds and supports microscope slides.

13. **Stage clips:** Hold slides in place on the stage.

14. **Mechanical stage** (optional): A stage with stage clips that can be positioned using mechanical stage control knobs. Does your microscope have a mechanical stage? _____

15. **Mechanical stage control knobs** (optional): Two knobs usually located below the stage. One knob controls forward/reverse movement of the stage clips, and the other controls right/left movement.

## Focusing the Microscope—Lowest Power

1. Turn the nosepiece so that the *lowest* power lens is in straight alignment over the stage.

2. Always begin focusing with the *lowest* power objective lens (4× [scanning] or 10× [low power]).

3. With the coarse-adjustment knob, lower the stage (or raise the objectives) until it stops.

4. Place a slide of the letter *e* on the stage, and stabilize it with the clips. (If your microscope has a mechanical stage, pinch the spring of the slide arms on the stage, and insert the slide.) Center the *e* as best you can on the stage or use the two control knobs located below the stage (if your microscope has a mechanical stage) to center the *e*.

5. Again, be sure that the lowest power objective is in place. Then, as you look from the side, decrease the distance between the stage and the tip of the objective lens until the lens comes to an automatic stop or is no closer than 3 mm above the slide.

6. While looking into the eyepiece, rotate the diaphragm (or diaphragm control lever) to give the maximum amount of light.

7. Using the coarse-adjustment knob, slowly increase the distance between the stage and the objective lens until the object—in this case, the letter *e*—comes into view, or focus.

8. Once the object is seen, you may need to adjust the amount of light. To increase or decrease the contrast, rotate the diaphragm slightly.

9. Use the fine-adjustment knob to sharpen the focus if necessary.

10. Practice having both eyes open when looking through the eyepieces, as this greatly reduces eyestrain.

## Inversion

Inversion refers to the fact that a microscopic image is upside down and reversed.

### Observation: Inversion

1. In space 1 provided here, draw the letter *e* as it appears on the slide (with the unaided eye, not looking through the eyepiece).

| 1. | 2. |
|---|---|
|  |  |

2. In space 2, draw the letter *e* as it appears when you look through the eyepiece.

3. What differences do you notice? _____

4. Move the slide to the right. Which way does the image appear to move? _____
   Explain. _____

## Focusing the Microscope—Higher Powers

Compound light microscopes are **parfocal**—that is, once the object is in focus with the lowest power, it should also be almost in focus with the higher power.

1. Bring the object into focus under the lowest power by following the instructions in the previous section.

2. Make sure that the letter *e* is centered in the field of the lowest objective.

3. Move to the next higher objective (low power [10×] or high power [40×]) by turning the nosepiece until you hear it click into place. Do not change the focus; parfocal microscope objectives will not hit normal slides when changing the focus if the lowest objective is initially in focus. (If you are on low power [10×], proceed to high power [40×] before going on to step 4.)

4. If any adjustment is needed, use only the *fine*-adjustment knob. (*Note:* Always use only the fine-adjustment knob with high power.) On your drawing of the letter *e*, draw a circle around the portion of the letter that you are now seeing with high-power magnification.

5. When you have finished your observations of this slide (or any slide), rotate the nosepiece until the lowest power objective clicks into place, lower the stage (or raise the nosepiece), and then remove the slide.

## Total Magnification

Total magnification is calculated by multiplying the magnification of the ocular lens (eyepiece) by the magnification of the objective lens.

### Observation: Total Magnification

Calculate total magnification figures for your microscope, and record your findings in Table 2.3.

| Table 2.3 Total Magnification | | | |
|---|---|---|---|
| Objective | Ocular Lens | Objective Lens | Total Magnification |
| Scanning power (if present) | | | |
| Low power | | | |
| High power | | | |
| Oil immersion (if present) | | | |

## 2.5 Microscopic Observations

When a specimen is prepared for observation, the object should always be viewed as a **wet mount.** A wet mount is prepared by placing a drop of liquid specimen on a slide or, if the specimen is dry, by placing it directly on the slide and adding a drop of water or stain. The mount is then covered with a coverslip, as illustrated in Figure 2.6. Dry the bottom of your slide before placing it on the stage.

**Figure 2.6** **Preparation of a wet mount.**

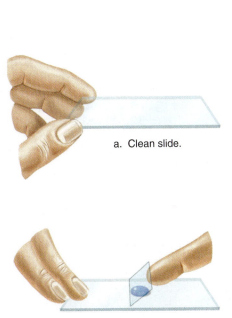

a. Clean slide.

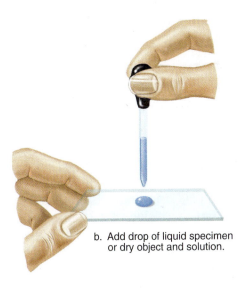

b. Add drop of liquid specimen or dry object and solution.

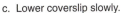

c. Lower coverslip slowly.

d. View wet mount.

## Human Epithelial Cells (Animal Cell)

Epithelial cells cover the body's surface and line its cavities.

> **Caution:** **Methylene blue** Avoid ingestion, inhalation, and contact with skin, eyes, and mucous membranes. Exercise care in using this chemical. If any should spill on your skin, wash the area with mild soap and water. Methylene blue will also stain clothing. Follow your instructor's directions for disposal of this chemical.

### Observation: Human Epithelial Cells

1. Obtain a prepared slide, or make your own as follows:
   a. Obtain a prepackaged flat toothpick (or sanitize one with alcohol or alcohol swabs).
   b. Gently scrape the inside of your cheek with the toothpick, and place the scrapings on a clean, dry slide. Discard used toothpicks in the biohazard waste container provided.
   c. Add a drop of very weak *methylene blue* or *iodine solution*, and cover with a coverslip.
2. Observe under the microscope.
3. Locate the nucleus (the central round body), the cytoplasm, and the plasma membrane (outer cell boundary). *Label Figure 2.7.*
4. Because your epithelial slides are biohazardous, they must be disposed of as indicated by your instructor.

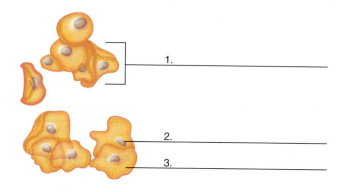

**Figure 2.7** **Cheek epithelial cells.**
Label the nucleus, the cytoplasm, and the plasma membrane.

1. _____
2. _____
3. _____

## Onion Epidermal Cells (Plant Cell)

Epidermal cells cover the surfaces of plant organs such as leaves. The bulb of an onion is made up of fleshy leaves.

### Observation: Onion Epidermal Cells

1. With a scalpel, strip a small, thin, transparent layer of cells from the inside of a fresh onion leaf.
2. Place it gently on a clean, dry slide, and add a drop of *iodine solution* (or *methylene blue*). Cover with a coverslip.
3. Observe under the microscope.
4. Locate the cell wall and the nucleus. *Label Figure 2.8.*
5. Note some obvious differences between the human cheek cells and the onion cells, and list them in Table 2.4.

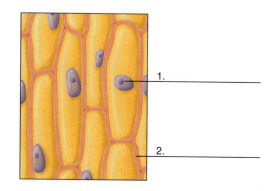

**Figure 2.8** **Onion epidermal cells.**
Label the cell wall and the nucleus.

1. _____
2. _____

| Table 2.4 | Differences Between Human Epithelial and Onion Epidermal Cells | |
|---|---|---|
| **Differences** | **Human Epithelial Cells (Cheek)** | **Onion Epidermal Cells** |
| | | |
| | | |
| Shape (sketch) | | |
| | | |

## Euglena

Examination of *Euglena* (a unicellular organism with a flagellum to facilitate movement) will test your ability to observe objects with the microscope and to control illumination to heighten contrast.

### Observation: Euglena

1. Make a wet mount of *Euglena* by using a drop of a *Euglena* culture and adding a drop of Protoslo® (methyl cellulose solution) onto a slide. The Protoslo slows the organism's swimming.
2. Mix thoroughly with a toothpick, and add a coverslip.
3. Scan the slide for *Euglena:* Start at the upper left-hand corner, and move the slide forward and back as you work across the slide from left to right. The *Euglena* may be at the edge of the slide because they show an aversion to Protoslo. Use Figure 2.9 to help identify the structural details of *Euglena*.
4. Experiment by using scanning, low-power, and high-power objective lenses, by focusing up and down with the fine-adjustment knob, and by adjusting the light so that it is not too bright.
5. Compare your *Euglena* specimens with Figure 2.9. Can you see any cellular contents? Describe what you can see. _____

_____

**Figure 2.9** *Euglena.*
*Euglena* is a unicellular, flagellated organism.

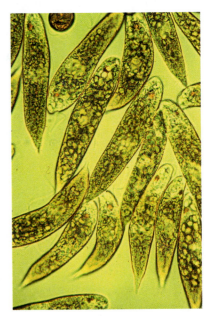

_____  **1.** 11 mm equals how many cm?

_____  **2.** 950 mm equals how many m?

_____  **3.** 2.1 liters equals how many ml?

_____  **4.** If a wooden block measures 3 cm × 3 cm × 3 cm, how many ml would you expect it to displace in a beaker of water?

_____  **5.** Which type of microscope would you use to view a *Euglena* swimming in pond water?

_____  **6.** What are the ocular lenses?

_____  **7.** Which objective always should be in place, both when beginning to use the microscope and when putting it away?

_____  **8.** A total magnification of 100× requires the use of the 10× ocular lens with which objective?

_____  **9.** Which part of a microscope regulates the amount of light?

_____  **10.** What word is used to indicate that if the object is in focus at low power, it will also be in focus at high power?

_____  **11.** What adjustment knob is used with high power?

_____  **12.** If a *Euglena* is swimming to the left, which way should you move your slide to keep it in view?

_____  **13.** What is the final item placed on a wet mount before viewing with a light microscope?

_____  **14.** What type of object do you study with a binocular dissecting microscope?

_____  **15.** Why is a binocular dissecting microscope also called a stereomicroscope?

### Thought Questions

**16.** A virus is 50 nm in size. Which type of microscope should be used to view it? Why?

**17.** Why is locating an object more difficult if you start with the high-power objective rather than with the low-power objective?

# 3

# Cell Structure and Function

## Learning Objectives

**3.1 Prokaryotic Versus Eukaryotic Cells**
- Distinguish between prokaryotic and eukaryotic cells by description and examples.

**3.2 Animal Cell and Plant Cell Structure**
- Label an animal cell diagram and state a function for the structures labeled.
- Label a plant cell diagram and state a function for the structures labeled.

**3.3 Diffusion**
- Define and describe the process of diffusion.
- Predict which substances will or will not diffuse across a plasma membrane.

**3.4 Osmosis**
- Define isotonic, hypertonic, and hypotonic solutions.
- Predict the effect of different tonicities on animal (e.g., red blood) cells and on plant (e.g., *Elodea*) cells.

**3.5 pH and Cells**
- Predict the pH before and after the addition of an acid to nonbuffered and buffered solutions.

## Introduction

Molecules are not alive—the basic units of life are cells. The **cell theory** states that all living things are composed of cells and that cells come only from other cells. Some organisms, such as *Euglena*, which you observed in Laboratory 2, are unicellular, but multicellular organisms are composed of many cells. While we are accustomed to considering the heart, the liver, or the intestines as enabling the human body to function, it is actually cells that do the work of these organs.

Figure 3.1 is a human cheek epithelial cell as viewed by an ordinary compound light microscope available in general biology laboratories. It shows that the content of a cell, called the **cytoplasm,** is bounded by a **plasma membrane.** The plasma membrane regulates the movement of molecules into and out of the cytoplasm. In this lab, we will study how the passage of water into a cell depends on the difference in concentration of solutes (particles) between the cytoplasm and the surrounding medium or solution. The well-being of cells also depends upon the pH of the solution surrounding them. We will see how a buffer can maintain the pH within a narrow range and how buffers within cells can protect them against damaging pH changes.

Because a photomicrograph shows only a minimal amount of detail, it is necessary to turn to the electron microscope to study the contents of a cell in greater depth. The models of plant and animal cells available in the laboratory today are based on electron micrographs.

**Figure 3.1** **Photomicrograph of an epithelial cell.**

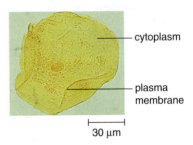

cytoplasm

plasma membrane

30 μm

# 3.1 Prokaryotic Versus Eukaryotic Cells

All living cells are classified as either prokaryotic or eukaryotic. One of the basic differences between the two types is that prokaryotic cells do not contain nuclei (*pro,* "before"; *karyote,* "nucleus"), while eukaryotic cells do contain nuclei (*eu,* "true"; *karyote,* "nucleus"). Only bacteria (including cyanobacteria) and archaea are prokaryotes; all other organisms are eukaryotes.

Prokaryotes also don't have the organelles found in eukaryotic cells (Fig. 3.2). **Organelles** are small, membranous bodies, each with a specific structure and function. Prokaryotes have **cytoplasm,** the material bounded by a plasma membrane and cell wall. The cytoplasm contains ribosomes, small granules that coordinate the synthesis of proteins; thylakoids (only in cyanobacteria) that participate in photosynthesis; and innumerable enzymes. Prokaryotes also have a nucleoid, a region in the bacterial cell interior in which the DNA is physically organized but not enclosed by a membrane.

## Figure 3.2 Prokaryotic cells.

*a.* Nonphotosynthetic bacterium. *b.* Cyanobacterium, a photosynthetic bacterium, formerly called a blue-green alga.

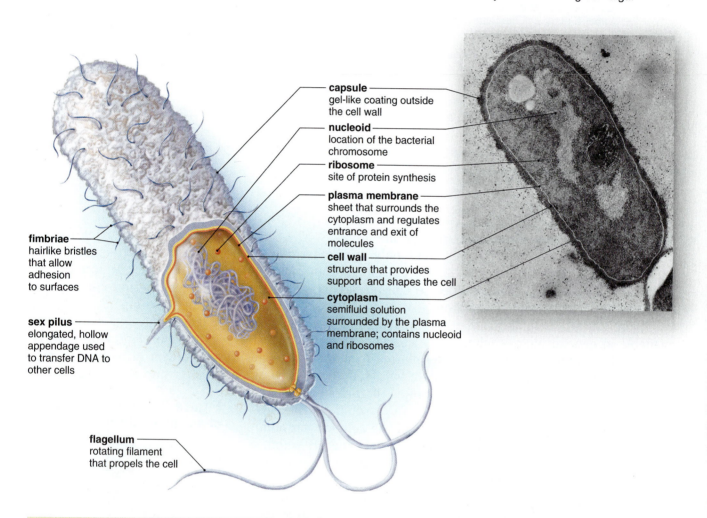

**capsule**
gel-like coating outside the cell wall

**nucleoid**
location of the bacterial chromosome

**ribosome**
site of protein synthesis

**plasma membrane**
sheet that surrounds the cytoplasm and regulates entrance and exit of molecules

**cell wall**
structure that provides support and shapes the cell

**cytoplasm**
semifluid solution surrounded by the plasma membrane; contains nucleoid and ribosomes

**fimbriae**
hairlike bristles that allow adhesion to surfaces

**sex pilus**
elongated, hollow appendage used to transfer DNA to other cells

**flagellum**
rotating filament that propels the cell

## Observation: Prokaryotic/Eukaryotic Cells

Two microscope slides will show you the main difference between prokaryotic and eukaryotic cells.

1. Examine a prepared slide of a prokaryote. Notice that there are no nuclei in these cells. Sketch this cell.
2. Examine a prepared slide of cuboidal cells from a human kidney. Notice that you can make out a nucleus. Sketch this cell next to the prokaryotic cell.

# 3.2 Animal Cell and Plant Cell Structure

Table 3.1 lists the structures found in animal and plant cells. The **nucleus** in a eukaryotic cell is bounded by a **nuclear envelope** and contains **nucleoplasm.** The *cytoplasm,* found between the plasma membrane and the nucleus, consists of a background fluid and the organelles. Many **organelles** are membranous, such as the nucleolus, endoplasmic reticulum, Golgi apparatus, vacuoles and vesicles, lysosomes, peroxisome, mitochondrion, and chloroplast. *Study Table 3.1 to determine structures that are unique to plant cells and unique to animal cells, and write them below the examples given.*

| Table 3.1 | Eukaryotic Structures in Animal Cells and Plant Cells | |
|---|---|---|
| **Name** | **Composition** | **Function** |
| Cell wall* | Contains cellulose fibrils | Provides support and protection |
| Plasma membrane | Phospholipid bilayer with embedded proteins | Outer cell surface that regulates entrance and exit of molecules |
| Nucleus | Enclosed by nuclear envelope; contains chromatin (threads of DNA and protein) and nucleolus (produces ribosomes) | Storage of genetic information; synthesis of DNA and RNA |
| Nucleolus | Concentrated area of chromatin, RNA, and proteins | Produces subunits of ribosomes |
| Ribosome | Protein and RNA in two subunits | Carries out protein synthesis |
| Endoplasmic reticulum (ER) | Membranous, flattened channels and tubular canals; rough ER and smooth ER | Synthesis and/or modification of proteins and other substances; transport by vesicle formation |
| Rough ER | Studded with ribosomes | Protein synthesis |
| Smooth ER | Lacks ribosomes | Synthesis of lipid molecules |
| Golgi apparatus | Stack of membranous saccules | Processes, packages, and distributes proteins and lipids |
| Vacuole and vesicle | Membrane-bounded sac | Storage of substances |
| Lysosome | Vesicle containing hydrolytic enzymes | Digests macromolecules and cell parts |
| Peroxisome | Vesicle containing specific enzymes | Breaks down fatty acids and converts resulting hydrogen peroxide to water; various other functions |
| Mitochondrion | Membranous thylakoids bounded by an outer membrane | Carries out cellular respiration, producing ATP molecules |
| Chloroplast* | Membranous cristae bounded by two membranes | Carries out photosynthesis, producing sugars |
| Cytoskeleton | Microtubules, intermediate filaments, actin filaments | Maintains cell shape and assists movement of cell parts |
| Cilia and flagella | 9 + 2 pattern of microtubules | Movement of cell |
| Centrioles** | 9 + 0 pattern of microtubules | Unknown function |

*Plant cells only

**Animal cells only

Unique structures:

**Plant Cells**
1. Large central vacuole
2. _____
3. _____

**Animal Cells**
Small vacuoles
_____

## Animal Cell Structure

With the help of Table 3.1, give a function for each of these structures, and label Figure 3.3. See also Figure 4.7 of the text.

| Structure | Function |
|---|---|
| Plasma membrane | |
| Nucleus | |
| Nucleolus | |
| Ribosomes | |
| Endoplasmic reticulum | |
|     Rough ER | |
|     Smooth ER | |
| Golgi apparatus | |
| Vacuole and vesicle | |
| Lysosome | |
| Mitochondrion | |
| Centriole | |
| Cilia and flagella (not shown) | |

**Figure 3.3** Animal cell structure.

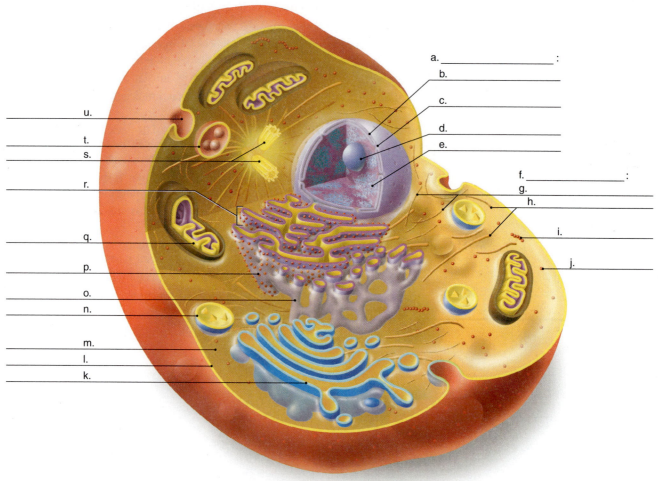

## Plant Cell Structure

With the help of Table 3.1, give a function for these structures unique to plant cells, and label Figure 3.4. See also Figure 4.8 in the text.

| Structure | Function |
|---|---|
| Cell wall | |
| Central vacuole, large | |
| Chloroplasts | |

**Figure 3.4** Plant cell structure.

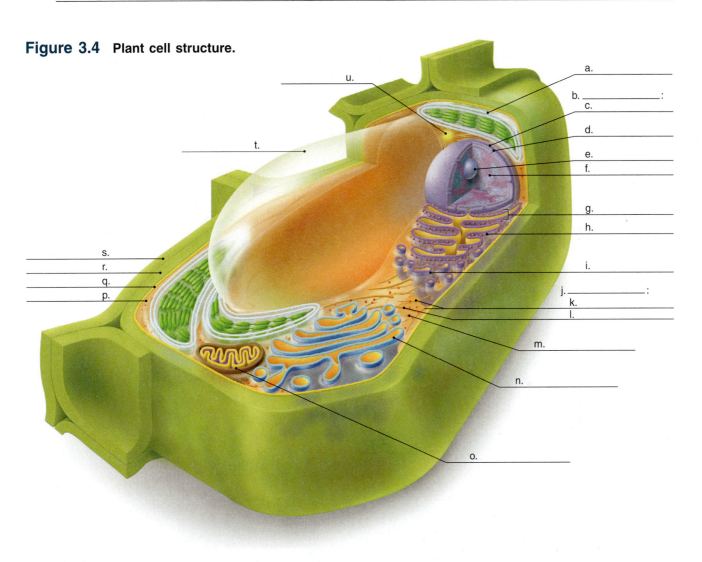

---

### Observation: Plant Cell Structure

1. Prepare a wet mount of a small piece of young *Elodea* leaf in fresh water. *Elodea* is a multi-cellular, eukaryotic plant found in freshwater ponds and lakes.
2. Have the drop of *water* ready on your slide so that the leaf does not dry out, even for a few seconds. Take care that the leaf is mounted with its top side up.
3. Examine the slide using low power, focusing sharply on the leaf surface.

4. Select a cell with numerous chloroplasts for further study, and switch to high power.

5. Carefully focus on the side and end walls of the cell. The chloroplasts appear to be only along the sides of the cell because the large, fluid-filled, membrane-bounded central vacuole pushes the cytoplasm against the cell walls (Fig. 3.5a). Then focus on the surface and notice an even distribution of chloroplasts (Fig. 3.5b).

6. Can you locate the cell nucleus? _____ It may be hidden by the chloroplasts, but when visible, it appears as a faint, grey lump on one side of the cell.

7. Can you detect movement of chloroplasts in this cell or any other cell? _____ The chloroplasts are not moving under their own power but are being carried by a streaming of the nearly invisible cytoplasm.

8. Save your slide for use later in this laboratory.

**Figure 3.5** *Elodea* cell structure.

central vacuole

cell wall

chloroplasts

cytoplasm

5 μm

a. Middle of the cell, with chloroplasts visible around the perimeter and not in the center, which is occupied by a membrane-bounded, fluid-filled, central vacuole.

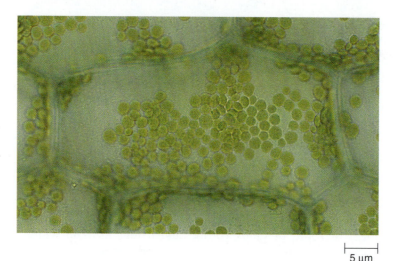

5 μm

b. Upper surface of cells, showing chloroplasts in the middle, as well as around the perimeter.

# 3.3 Diffusion

**Diffusion** is the movement of molecules from a higher to a lower concentration until equilibrium is achieved and the molecules are distributed equally (Fig. 3.6). At this point, molecules may still be moving back and forth, but there is no net movement in any one direction.

**Figure 3.6 Process of diffusion.**
Diffusion is apparent when dye molecules have equally dispersed.

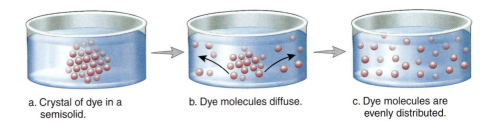

a. Crystal of dye in a semisolid.

b. Dye molecules diffuse.

c. Dye molecules are evenly distributed.

Diffusion is a general phenomenon in the environment. The speed of diffusion is dependent on such factors as the temperature, the size of the molecule, and the type of medium.

## Experimental Procedure: Diffusion

**Caution:** **Potassium permanganate ($KMnO_4$)** $KMnO_4$ is highly poisonous and is a strong oxidizer. Avoid contact with skin and eyes and with combustible materials. If spillage occurs, wash all surfaces thoroughly. $KMnO_4$ will also stain clothing.

### Diffusion Through a Semisolid

1. Observe a petri dish containing 1.5% gelatin (or agar) to which potassium permanganate ($KMnO_4$) was added in the center depression at the beginning of the lab.
2. Obtain *time zero* from your instructor, and record *time zero* and the *final time* (now) in Table 3.2. Calculate the length of time in hours and minutes. Convert the time to hours: _____ hr.
3. Using a ruler placed over the petri dish, measure (in mm) the movement of color from the center of the depression outward in one direction: _____ mm.
4. Calculate the speed of diffusion: _____ mm/hr.
5. Record all data in Table 3.2.

### Diffusion Through a Liquid

1. Add enough water to cover the bottom of a glass petri dish.
2. Place the petri dish over a thin, flat ruler.
3. With tweezers, add a crystal of potassium permanganate ($KMnO_4$) directly over a millimeter measurement line. Note the *time zero* in Table 3.2.
4. After 10 minutes, note the distance the color has moved. Record the *final time, length of time,* and *distance moved* in Table 3.2.
5. Multiply the length of time and the distance moved by 6 to calculate the *speed of diffusion:* _____ mm/hr. Record in Table 3.2.

### Diffusion Through Air

1. Measure the distance from a spot designated by your instructor to your laboratory work area today. Record this distance in the fifth column of Table 3.2.
2. Record *time zero* in Table 3.2 when a perfume or similar substance is released into the air.

3. Note the time when you can smell the perfume. Record this as the *final time* in Table 3.2. Calculate the *length of time* since the perfume was released, and record it in Table 3.2.

4. Calculate the speed of diffusion: _____ mm/hr. Record in Table 3.2.

| Table 3.2 | Speed of Diffusion | | | | |
|-----------|-----------|------------|-------------------|---------------------|-------------------------|
| Medium | Time Zero | Final Time | Length of Time (hr) | Distance Moved (mm) | Speed of Diffusion (mm/hr) |
| Semisolid | | | | | |
| Liquid | | | | | |
| Air | | | | | |

## Conclusions

- In which experiment was diffusion the fastest? _____
- What accounts for the difference in speed? _____

## Diffusion Across the Plasma Membrane

Some molecules can diffuse across a plasma membrane, and some cannot. In general, small, non-charged molecules can cross a membrane by simple diffusion, but large molecules cannot diffuse across a membrane. The dialysis tube membrane in the experimental procedure simulates a plasma membrane.

### Experimental Procedure: Diffusion Across Plasma Membrane

1. Cut a piece of dialysis tubing approximately 40 cm (approximately 16 inches) long. Soak the tubing in water until it is soft and pliable.
2. Close one end of the dialysis tubing with two knots.
3. Fill the bag halfway with *glucose solution*, i.e., glucose dissolved in water.
4. Add 4 full droppers of *starch solution* to the bag, i.e., starch dissolved in water.
5. Hold the open end while you mix the contents of the dialysis bag. Rinse off the outside of the bag with *distilled water.*
6. Fill a beaker 2/3 full with *distilled water.*
7. Add droppers of *iodine solution* (IKI) to the water in the beaker until an amber (tealike) color is apparent.
8. Record the color of the solution in the beaker in Table 3.3.

**Figure 3.7** **Placement of dialysis bag in water containing iodine.**

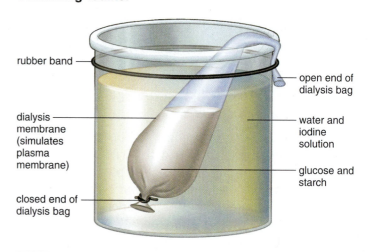

rubber band

open end of dialysis bag

dialysis membrane (simulates plasma membrane)

water and iodine solution

glucose and starch

closed end of dialysis bag

**Caution:** **Benedict's reagent** Exercise care in using this chemical. It is highly corrosive. If any should spill on your skin, wash the area with mild soap and water. Follow your instructor's directions for its disposal.

9. Place the bag in the beaker with the open end hanging over the edge. Secure the open end of the bag to the beaker with a rubber band as shown (Fig. 3.7). Make sure the contents do not spill into the beaker.

After about five minutes, at the end of the experiment,

10. You will note a color change. Record the color of the bag contents in Table 3.3.
11. Mark off a test tube at 1 cm and 3 cm.
12. Withdraw solution from near the bag and at the bottom of the beaker for testing with Benedict's reagent. Fill the test tube to the first mark with this solution. Add *Benedict's reagent* to the 3 cm mark. Heat in a boiling water bath for 5–10 minutes, observe any color change, and record your results as positive or negative in Table 3.3. (Optional use of glucose test strip: Dip glucose test strip into beaker. Compare stick with chart provided by instructor.)
13. Remove the dialysis bag from the beaker. Dispose of it and dispose of Benedict's reagent solution in the manner directed by your instructor.

| Table 3.3 | Diffusion Across Plasma Membrane | | | | |
|---|---|---|---|---|---|
| | At Start of Experiment | | At End of Experiment | | |
| | Contents | Color | Color | Benedict's Test (+) or (−) | Conclusion |
| Bag | Glucose Starch | | NA | | |
| Beaker | Water Iodine | | | | |

## Conclusions

- Based on the color change noted in the bag, conclude what solute diffused across the dialysis membrane from the beaker to the bag, and record your conclusion in Table 3.3.
- From the results of the Benedict's test on the beaker contents, conclude what solute diffused across the dialysis membrane from the bag to the beaker, and record your conclusion in Table 3.3.
- Which solute did not diffuse across the dialysis membrane from the bag to the beaker? _____ _____ Explain. _____

## Laboratory Notes

_____
_____
_____
_____
_____
_____
_____
_____
_____
_____

# 3.4 Osmosis

**Osmosis** is the diffusion of water across the plasma membrane of a cell. Just like any other molecule, water follows its concentration gradient and moves from the area of higher concentration to the area of lower concentration.

## Experimental Procedure: Osmosis

To demonstrate osmosis, a thistle tube is covered with a membrane at its lower opening and partially filled with 50% corn syrup (a polysaccharide) or a similar substance. The whole apparatus is placed in a beaker containing distilled water (Fig. 3.8). The water concentration in the beaker is 100%. Water molecules can move freely between the thistle tube and the beaker.

1. Note the level of liquid in the thistle tube, and measure how far it travels in 10 minutes:

   _____ mm.

2. Calculate the speed of osmosis under these conditions: _____ mm/hr.

### Conclusions

- In which direction was there a net movement of water? _____
  Explain what is meant by "net movement" after examining the arrows in Figure 3.8b.

  _____

- If the sugar molecules in corn syrup moved from the thistle tube to the beaker, would there have been a net movement of water into the thistle tube? _____ Why wouldn't large sugar molecules be able to move across the membrane from the thistle tube to the beaker?

  _____

- Explain why the water level in the thistle tube rose: In terms of solvent concentration, water moved from the area of _____ water concentration to the area of _____ water concentration across a differentially permeable membrane.

## Figure 3.8  Osmosis demonstration.

*a.* A thistle tube, covered at the broad end by a differentially permeable membrane, contains a 50% corn syrup solution. The beaker contains distilled water. *b.* The polysaccharides in corn syrup are unable to pass through the membrane, but the water passes through in both directions. There is a net movement of water toward the inside of the thistle tube, where there is a lower percentage of water molecules. *c.* Due to the incoming water molecules, the level of the solution rises in the thistle tube.

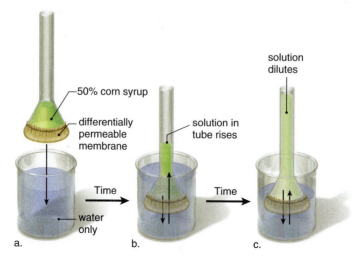

## Tonicity

**Tonicity** is the relative concentration of solute (particles), and therefore also of solvent (water), outside the cell compared with inside the cell.

- An **isotonic solution** has the same concentration of solute (and therefore of water) as the cell. When cells are placed in an isotonic solution, there is no net movement of water.
- A **hypertonic solution** has a higher solute (therefore, lower water) concentration than the cell. When cells are placed in a hypertonic solution, water moves out of the cell into the solution.
- A **hypotonic solution** has a lower solute (therefore, higher water) concentration than the cell. When cells are placed in a hypotonic solution, water moves from the solution into the cell.

### Experimental Procedure: Potato Strips

1. Cut two strips of potato, each about 7 cm long and 1.5 cm wide.
2. Label two test tubes 1 and 2. Place one *potato strip* in each tube.
3. Fill tube 1 with *water* to cover the potato strip.
4. Fill tube 2 with 10% *sodium chloride* (NaCl) to cover the potato strip.
5. After 1 hour, remove the potato strips from the test tubes and place them on a paper towel. Observe each strip for limpness (water loss) or stiffness (water gain). Which tube has

   the limp potato strip? _____ Why did water diffuse out of the potato strip in this tube?

   _____

   Which tube has the stiff potato strip? _____ Why did water diffuse into the potato strip in

   this tube? _____

### Red Blood Cells (Animal Cells)

A solution of 0.9% NaCl is isotonic to red blood cells. In such a solution, red blood cells maintain their normal appearance (Fig. 3.9a). A solution greater than 0.9% NaCl is hypertonic to red blood cells. In such a solution, the cells shrivel up, a process called **crenation** (Fig. 3.9b). A solution of less than 0.9% NaCl is hypotonic to red blood cells. In such a solution, the cells swell to bursting, a process called **hemolysis** (Fig. 3.9c).

**Figure 3.9** Tonicity and red blood cells.

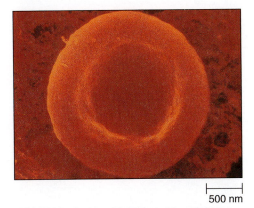

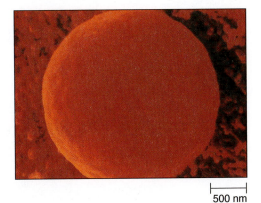

500 nm    500 nm    500 nm

**a. Isotonic solution.** Red blood cell has normal appearance due to no net loss or gain of water.

**b. Hypertonic solution.** Red blood cell shrivels due to loss of water.

**c. Hypotonic solution.** Red blood cell fills to bursting due to gain of water.

## Experimental Procedure: Demonstration of Tonicity in Red Blood Cells

> **Caution:** Do not remove the stoppers of test tubes during this procedure.

Three stoppered test tubes on display have the following contents:

Tube 1: 0.9% NaCl plus a few drops of whole sheep blood
Tube 2: 10% NaCl plus a few drops of whole sheep blood
Tube 3: Distilled water and a few drops of whole sheep blood

1. In the second column of Table 3.4, record the tonicity of each tube in relation to red blood cells.
2. Hold each tube in front of one of the pages of your lab manual. Determine whether you can see the print on the page through the tube. Record your findings in the third column of Table 3.4.

| Table 3.4 | Tonicity and Print Visibility | | |
|---|---|---|---|
| Tube | Tonicity | Print Visibility | Explanation |
| 1 | | | |
| 2 | | | |
| 3 | | | |

### Conclusion

- Explain in the fourth column of Table 3.4 why you can or cannot see the print.

### Elodea (Plant Cells)

When plant cells are in a hypotonic solution, the large central vacuole gains water and exerts pressure, called **turgor pressure.** The cytoplasm, including the chloroplasts, is pushed up against the cell wall (Fig. 3.10*a*).

When plant cells are in a hypertonic solution, the central vacuole loses water, and the cytoplasm, including the chloroplasts, pulls away from the cell wall. This is called **plasmolysis** (Fig. 3.10*b*).

## Experimental Procedure: Elodea Cells

### Hypotonic Solution

1. If possible, use the *Elodea* slide you prepared earlier in this laboratory. If not, prepare a new wet mount of a small *Elodea* leaf using fresh water.
2. After several minutes, focus on the surface of the cells, and compare your slide with Figure 3.10*a*.
3. Complete the portion of Table 3.5 that pertains to a hypotonic solution.

### Hypertonic Solution

1. Prepare a new wet mount of a small *Elodea* leaf using a 10% NaCl solution.
2. After several minutes, focus on the surface of the cells, and compare your slide with Figure 3.10*b*.
3. Complete the portion of Table 3.5 that pertains to a hypertonic solution.

| Table 3.5 | Effect of Tonicity on *Elodea* Cells | |
|---|---|---|
| **Tonicity** | **Appearance of Cells** | **Due to (scientific term)** |
| Hypotonic | | |
| Hypertonic | | |

### Conclusions

- In a hypotonic solution, the large central vacuole of plant cells exerts _____ pressure, and the chloroplasts are seen _____ the cell wall.
- In a hypertonic solution, the central vacuole loses water, and the chloroplasts are seen _____ the cell wall.

### Figure 3.10  *Elodea* cells.

*a.* Surface view of cells in a hypotonic solution and longitudinal section diagram. The large central vacuole, filled with water, pushes the cytoplasm, including the chloroplasts, right up against the cell wall. *b.* Surface view of cells in a hypertonic solution and longitudinal section diagram. When the central vacuole loses water, cytoplasm, including the chloroplasts, piles up in the center of the cell because the cytoplasm has pulled away from the cell wall. (*a:* Magnification ×400)

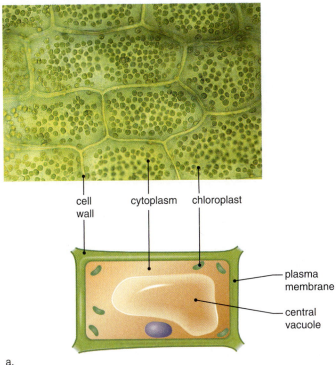

cell wall — cytoplasm — chloroplast — plasma membrane — central vacuole

a.

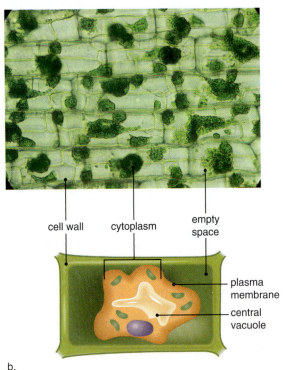

cell wall — cytoplasm — empty space — plasma membrane — central vacuole

b.

# 3.5 pH and Cells

The pH of a solution tells its hydrogen ion concentration [$H^+$]. The **pH scale** ranges from 0 to 14. A pH of 7 is neutral. A pH lower than 7 indicates that the solution is acidic (has more hydrogen ions than hydroxide ions), whereas a pH greater than 7 indicates that the solution is basic (has more hydroxide ions than hydrogen ions). A **buffer** is a system of chemicals that takes up excess hydrogen ions or hydroxide ions, as appropriate.

The concept of pH is important in biology because living organisms are very sensitive to hydrogen ion concentration. For example, in humans the pH of the blood must be maintained at about 7.4 or we become ill. All living things need to maintain the hydrogen ion concentration, or pH, at a constant level.

Why are cells and organisms buffered? _____

_____

## Experimental Procedure: pH and Cells

**Caution:** Hydrochloric acid (HCl) used to produce an acid pH is a strong, caustic acid. Exercise care in using this chemical. If any HCl spills on your skin, rinse immediately with clear water. Follow your instructor's directions for disposal of tubes that contain HCl.

1. Label three test tubes, and fill them to the halfway mark as follows: tube 1: *water;* tube 2: *buffer* (inorganic) solution; and tube 3: *simulated cytoplasm* (buffered protein solution).
2. Use pH paper to determine the pH of each tube. Dip the end of a stirring rod into the solution, and then touch the stirring rod to a 5 cm strip of pH paper. Read the current pH by matching the color observed with the color code on the pH paper package. Record your results in the "pH Before Acid" column in Table 3.6.
3. Add one drop of 0.1 N hydrochloric acid (HCl) to each tube. Shake or swirl. Use pH paper as in step 2 to determine the new pH of each solution. (Do as many pH determinations as possible on your pH paper before using a clean strip.) Record your results in the "pH After Acid" column in Table 3.6.

| Table 3.6 | pH and Cells | | | |
|-----------|--------------|--------------|-------------|-------------|
| Tube | Contents | pH Before Acid | pH After Acid | Explanation |
| 1 | Water | | | |
| 2 | Buffer | | | |
| 3 | Cytoplasm | | | |

## Conclusion

- Use the information in column two to explain your test results. Write your explanations in the last column of Table 3.6.

## Experimental Procedure: Buffer Strength

1. Pour the contents of tube 2 into a clean 50 ml beaker labeled "2." Pour the contents of tube 3 into a clean 50 ml beaker labeled "3."
2. Add 5M hydrochloric acid in 5-drop increments to beakers 2 and 3. Mix and test the pH after every 5 drops, and record these values in Table 3.7.
3. Record the patterns of pH changes on the graphs provided. Use the top graph for the pattern observed with the inorganic buffer and the bottom graph for the pattern observed with simulated cytoplasm.

| Table 3.7 | Drops of HCl and pH Values | |
|---|---|---|
| # Drops | pH | pH |
| | Beaker 2 (Buffer) | Beaker 3 (Cytoplasm) |
| 0 | | |
| 5 | | |
| 10 | | |
| 15 | | |
| 20 | | |
| 25 | | |
| 30 | | |
| 35 | | |
| 40 | | |
| 45 | | |
| 50 | | |

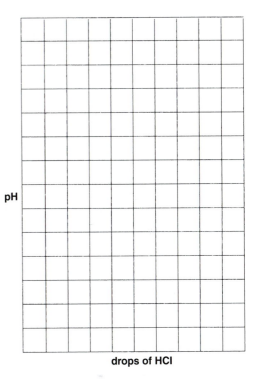

pH

drops of HCl

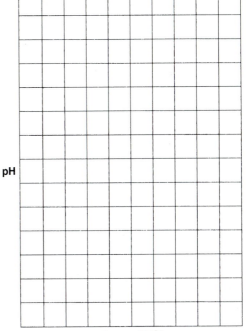

pH

drops of HCl

### Conclusion

- Do the two graphs have a similar pattern?

    _____

    Explain.

    _____

    _____

    _____

    _____

    _____

——————————————— 1. What is the name of the large, often central organelle in eukaryotic cells that contains chromosomes?

——————————————— 2. Prokaryotes have what structures necessary for protein synthesis?

——————————————— 3. What is the function of the nucleus?

——————————————— 4. What is the function of rough endoplasmic reticulum?

——————————————— 5. Which organelle carries on intracellular digestion?

——————————————— 6. Name a structure present in an animal cell but not in a plant cell.

——————————————— 7. Name a structure present in a plant cell but not in an animal cell.

——————————————— 8. What term describes the movement of molecules from an area of higher concentration to one of lower concentration?

——————————————— 9. What is the name for the movement of water across the plasma membrane?

——————————————— 10. In what direction does water move when cells are placed in a hypertonic solution?

——————————————— 11. Is 10% NaCl isotonic, hypertonic, or hypotonic to red blood cells?

——————————————— 12. What appearance will red blood cells have when placed in 9.0% NaCl?

——————————————— 13. What scientific term is used to refer to the condition of cells described in question 12?

——————————————— 14. What type of molecule prevents extensive changes in the pH of living organisms?

——————————————— 15. If acid is added to water, does the pH increase or decrease?

——————————————— 16. What is a pH of 7 called?

## Thought Questions

17. The police are trying to determine whether material removed from a crime scene is plant or animal matter. What would you suggest they look for?

18. Your grandmother asks you to fertilize her favorite plant. Without reading the directions on the box, you pour some fertilizer into the pot and then water the plant. The next time you see your grandmother, she tells you the plant died. In terms of osmosis, explain what happened to the plant.

# 4

# Enzymes

## Introduction

The cell carries out many chemical reactions. All the chemical reactions that occur in a cell are collectively called **metabolism.** A possible chemical reaction can be indicated like this:

$$A + B \longrightarrow C + D$$
$$\text{reactants} \quad \text{products}$$

In all chemical reactions, the **reactants** are molecules that undergo a change, which results in the **products.** The arrow stands for the change that produced the product(s). The number of reactants and products can vary; in the one you are studying today, a single reactant breaks down to two products.

All the reactions that occur in a cell have an enzyme. **Enzymes** are organic catalysts that speed metabolic reactions. Because enzymes are specific and speed only one type of reaction, they are given names. In today's laboratory, you will be studying the action of the enzyme **catalase.** The reactants in an enzymatic chemical reaction are called **substrate(s).**

Enzymes are specific because they have a shape that accommodates the shape of their substrates as a key fits a lock. Enzymatic reactions can be indicated like this:

$$E + S \longrightarrow ES \longrightarrow E + P$$

In this reaction, E = enzyme, S = substrate, ES = enzyme-substrate complex, and P = product.

## Figure 4.1 Enzymatic action.

The enzyme and substrate come together, and the reaction occurs on the surface of the enzyme at the active site.

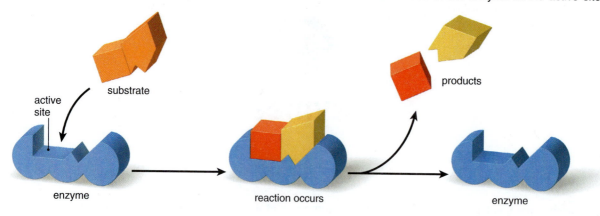

Two types of enzymatic reactions are common to cells. During degradation reactions (Fig. 4.1), the substrate is broken down to the product(s), and during synthesis reactions, the substrates are joined to form a product. Notice how the shape of the enzyme accommodates its substrate. The location where the enzyme and substrate form an enzyme-substrate complex is called the **active site** because the reaction occurs here.

At the end of the reaction, the product is released, and the enzyme can then combine with the substrate again. A cell needs only a small amount of an enzyme because enzymes are used over and over. Some enzymes have turnover rates well in excess of a million product molecules per minute.

All enzymes are complex proteins that generally act in an organism's closely controlled internal environment, where the temperature and pH remain within a rather narrow range. *Extremes* in pH or temperature may denature the enzyme by permanently altering its chemical structure. Even a small change in the protein's structure will change the enzyme's shape enough to prevent formation of the enzyme-substrate complex and thus keep the reaction from occurring. As more substrate molecules fill active sites, more product results per unit time. Therefore, in general, an increase in enzyme or substrate speeds enzymatic reactions. In this laboratory, you will test the effect of temperature, enzyme concentration, and pH on an enzymatic reaction.

# 4.1 Catalase Activity

In the Experimental Procedures that follows, you will be working with the enzyme catalase. Catalase is present in cells, where it speeds the breakdown of the toxic chemical hydrogen peroxide to water and oxygen:

$$2H_2O_2 \xrightarrow{\text{catalase}} 2H_2O + O_2$$

$$\underset{\text{hydrogen peroxide}}{} \qquad \underset{\text{water \quad oxygen}}{}$$

What is the reactant in this reaction? _____ What is the substrate for catalase? _____

What are the products in this reaction? _____ and _____ Bubbling occurs as the reaction

proceeds. Why? _____

## Experimental Procedure: Catalase Activity

With a wax pencil, label and mark three clean test tubes at the 1 cm and 5 cm levels.

Tube 1
1. Fill to the first mark with *catalase* buffered at pH 7.0, the optimum pH for catalase.
2. Fill to the second mark with *hydrogen peroxide.* Swirl well to mix, and wait at least 20 seconds for bubbling to develop.
3. Measure the height of the bubble column (in millimeters), and record your results in Table 4.1.

Tube 2
1. Fill to the first mark with *water.*
2. Fill to the second mark with *hydrogen peroxide.* Swirl well to mix, and wait at least 20 seconds.
3. Measure the height of the bubble column (in millimeters), and record your results in Table 4.1.

Tube 3
1. Fill to the first mark with *catalase.*
2. Fill to the second mark with *sucrose solution.* Swirl well to mix; wait 20 seconds.
3. Measure the height of the bubble column, and record your results in Table 4.1.

| Table 4.1 | Catalase Activity | | |
|---|---|---|---|
| Tube | Contents | Bubble Column Height | Explanation |
| 1 | Catalase Hydrogen peroxide | | |
| 2 | Water Hydrogen peroxide | | |
| 3 | Catalase Sucrose solution | | |

## Conclusions

- Which tube showed the bubbling you expected? _____ Conclude why this tube showed bubbling, and record your explanation in Table 4.1.

- Which tubes are controls? _____ If these tubes showed bubbling, what could you conclude about your procedure? _____ Complete Table 4.1.

- Enzymes are specific; they speed only a reaction that contains their substrate. Which tube exemplifies this characteristic of an enzyme? _____ Record your explanation in Table 4.1 for this tube.

## 4.2 Effect of Temperature on Enzyme Activity

In general, cold temperatures slow chemical reactions, and warm temperatures speed chemical reactions. Boiling, however, causes an enzyme to denature in a way that inactivates it.

### Experimental Procedure: Effect of Temperature

With a wax pencil, label and mark three clean test tubes at the 1 cm and 5 cm levels.

1. Fill each tube to the first mark with *catalase* buffered at pH 7.0, the optimum pH for catalase.
2. Place tube 1 in a refrigerator or cold water bath, tube 2 in an incubator or warm water bath, and tube 3 in a boiling water bath. Complete the second column in Table 4.2. Wait 15 minutes.
3. As soon as you remove the tubes one at a time from the refrigerator, incubator, and boiling water, fill to the second mark with *hydrogen peroxide*.
4. Swirl well to mix, and wait 20 seconds.
5. Measure the height of the bubble column (in millimeters) in each tube, and record your results in Table 4.2.
6. Plot your results in Figure 4.2. Put temperature (°C) on the X-axis and bubble column height (mm) on the Y-axis.

| Table 4.2 | Effect of Temperature | | |
|---|---|---|---|
| Tube | Temperature °C | Bubble Column Height (mm) | Explanation |
| 1  Refrigerator | | | |
| 2  Incubator | | | |
| 3  Boiling water | | | |

### Conclusions

- The amount of bubbling corresponds to the degree of enzyme activity. Explain in Table 4.2 the degree of enzyme activity per tube.

- What is your conclusion concerning the effect of temperature on enzyme activity?

_____

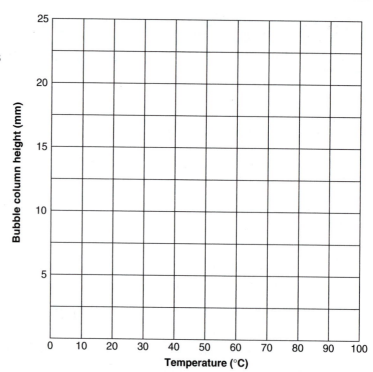

**Figure 4.2  Effect of temperature on enzyme activity.**

# 4.3 Effect of Concentration on Enzyme Activity

In general, a higher enzyme/substrate concentration results in faster enzyme activity—that is, the amount of product per unit time for any particular reaction will increase.

With a wax pencil, label three clean test tubes.

Tube 1
1. Mark this tube at the 1 cm and 5 cm levels.
2. Fill to the first mark with buffered *catalase* and to the second mark with *hydrogen peroxide.*
3. Swirl well to mix, and wait 20 seconds.
4. Measure the height of the bubble column (in millimeters), and record your results in Table 4.3.

Tube 2
1. Mark this tube at the 2 cm and 6 cm levels.
2. Fill to the first mark with buffered *catalase* and to the second mark with *hydrogen peroxide.*
3. Swirl well to mix, and wait 20 seconds.
4. Measure the height of the bubble column (in millimeters), and record your results in Table 4.3.

Tube 3
1. Mark this tube at the 3 cm and 7 cm levels.
2. Fill to the first mark with buffered *catalase* and to the second mark with *hydrogen peroxide.*
3. Swirl well to mix, and wait 20 seconds.
4. Measure the height of the bubble column (in millimeters), and record your results in Table 4.3.

| Table 4.3 | Effect of Enzyme Concentration | | |
|---|---|---|---|
| **Tube** | **Amount of Enzyme** | **Bubble Column Height (mm)** | **Explanation** |
| 1 | 1 cm | | |
| 2 | 2 cm | | |
| 3 | 3 cm | | |

## Conclusions

- The amount of bubbling corresponds to the degree of enzyme activity. Explain in Table 4.3 the degree of enzyme activity per tube.

- If unlimited time were allotted, would the results be the same in all tubes? _____
  Explain why or why not. _____

- Would you expect similar results if the substrate concentration were varied in the same manner as the enzyme concentration? _____ Why or why not? _____

## 4.4 Effect of pH on Enzyme Activity

Each enzyme has a pH at which the speed of the reaction is optimum (occurs best). Any higher or lower pH affects hydrogen bonding and the structure of the enzyme, leading to reduced activity.

### Experimental Procedure: Effect of pH

> **Caution:** Hydrochloric acid (HCl) used to produce an acid pH is a strong, caustic acid, and sodium hydroxide (NaOH) used to produce a basic pH is a strong, caustic base. Exercise care in using these chemicals, and follow your instructor's directions for disposal of tubes that contain these chemicals. If any acidic or basic solutions spill on your skin, rinse immediately with clear water.

With a wax pencil, label and mark three clean test tubes at the 1 cm, 3 cm, and 7 cm levels. Fill each tube to the 1 cm level with nonbuffered *catalase*.

**Tube 1**
1. Fill to the second mark with *water* adjusted to pH 3 by the addition of *HCl*.
2. Fill to the third mark with *hydrogen peroxide*. Wait one minute.
3. Swirl to mix, and wait 20 seconds.
4. Measure the height of the bubble column (in millimeters), and record your results in Table 4.4.

**Tube 2**
1. Fill to the second mark with *water* adjusted to pH 7.
2. Wait one minute. Fill to the third mark with *hydrogen peroxide*.
3. Swirl to mix, and wait 20 seconds.
4. Measure the height of the bubble column (in millimeters), and record your results in Table 4.4.

**Tube 3**
1. Fill to the second mark with *water* adjusted to pH 11 by the addition of *NaOH*.
2. Fill to the third mark with *hydrogen peroxide*. Wait one minute.
3. Swirl to mix, and wait 20 seconds.
4. Measure the height of the bubble column (in millimeters), and record your results in Table 4.4.
5. Plot your results in Figure 4.3. Put pH on the X-axis and column height (mm) on the Y-axis.

| Table 4.4 | | Effect of pH | |
|---|---|---|---|
| Tube | pH | Bubble Column Height (mm) | Explanation |
| 1 | 3 | | |
| 2 | 7 | | |
| 3 | 11 | | |

**Figure 4.3** Effect of pH on enzyme activity.

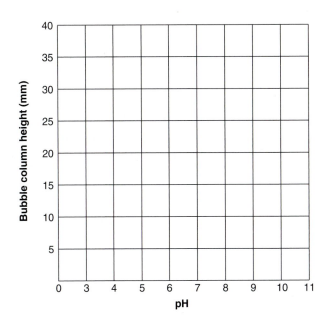

## Conclusions

- The amount of bubbling corresponds to the degree of enzyme activity. Explain in Table 4.4 the degree of enzyme activity per tube.
- The results of which tube in Table 4.1 could be used as a control for Table 4.4? _____

  Why could this tube be considered a control? _____

## Factors That Affect Enzyme Activity

In Table 4.5, summarize what you have learned about factors that affect the speed of an enzymatic reaction. For example, in general, what type of temperature promotes enzyme activity and what type inhibits enzyme activity? Answer similarly for enzyme/substrate concentration and pH.

| Table 4.5 | Factors That Affect Enzyme Activity | |
|---|---|---|
| **Factors** | **Promote Enzyme Activity** | **Inhibit Enzyme Activity** |
| Temperature | | |
| Enzyme/substrate concentration | | |
| pH | | |

## Conclusions

- Why does a warm temperature promote enzyme activity? _____
- Why does increasing enzyme concentration promote enzyme activity? _____
- Why does optimum pH promote enzyme activity? _____

————————— 1. In the representation of a chemical reaction, what does the arrow stand for?

————————— 2. The reactants in an enzymatic reaction are called what?

————————— 3. Where on an enzyme do substrates come together?

————————— 4. If an enzyme is boiled, what happens to the enzyme?

————————— 5. If an enzyme is warmed, what happens to its activity?

————————— 6. A control for the effect of a warm temperature would lack what factor?

————————— 7. If more enzyme is used, what happens to the amount of product per unit time?

————————— 8. What must be held constant when testing the effect of enzyme concentration on enzyme activity?

————————— 9. What product of the catalase reaction causes bubbling?

————————— 10. What is varied when testing the effect of pH on enzyme activity?

————————— 11. If the pH is unfavorable, what happens to enzyme activity?

————————— 12. What allowed you to measure the amount of enzyme activity?

## Thought Questions

13. Lipase is a digestive enzyme that digests fat droplets in the small intestine. Lipase requires a slightly basic pH, which the presence of $NaHCO_3$ provides. Indicate which of the following test tubes would show digestion following incubation, and explain why the others would not.

Tube 1  Water, fat droplets ————————————————————

Tube 2  Water, fat droplets, lipase ————————————————

Tube 3  Water, fat droplets, lipase, $NaHCO_3$ ————————————

Tube 4  Water, lipase, $NaHCO_3$ ————————————————

14. In what way does an enzyme speed its reaction? How does this explain why enzymes are specific?

# 5

# Cellular Respiration

**Introduction**
- Give the overall equation for fermentation and cellular respiration.
- Relate a cell's use of oxygen to each process.
- Explain the relationship between these processes and ATP molecules.

**5.1 Fermentation**
- Describe and explain the fermentation experiment.
- Relate the overall equation for fermentation to the fermentation experiment.
- State and explain the effects of food source on fermentation by yeast.

**5.2 Cellular Respiration**
- Describe and explain the cellular respiration experiment.
- Relate the overall equation for cellular respiration to the cellular respiration experiment.
- State and explain the effects of germination and nongermination of soybeans on the results of the cellular respiration experiment.

# Introduction

In this laboratory, you will study **fermentation** and **cellular respiration.** During fermentation, glucose is incompletely broken down, and much energy remains in the organic molecule that results. In cellular respiration, glucose is completely broken down to inorganic molecules. During fermentation, a small amount of chemical energy is converted to ATP molecules, and during cellular respiration much more chemical energy is converted to ATP molecules for use by the cell. ATP molecules are absolutely essential to cellular metabolism because they supply energy whenever a cell carries on activities, such as active transport, synthesis, muscle contraction, and nerve conduction.

In yeast, fermentation occurs in the following way:

$$C_6H_{12}O_6 \longrightarrow 2\ CO_2 + 2\ C_2H_5OH + 2\ ATP$$

glucose  →  carbon dioxide  ethanol

Yeast is used by the wine and beer industries to ferment the carbohydrates in fruits and grains to alcohol. In baking, the carbon dioxide given off from yeast fermentation causes bread to rise. Fermentation in animals and certain microbes produces lactic acid. This form of fermentation helps produce yogurt and many cheeses, as well as such products as sourdough breads, chocolate, and pickled foods.

In most organisms, cellular respiration occurs in the following way:

$$C_6H_{12}O_6 + 6\ O_2 \longrightarrow 6\ CO_2 + 6\ H_2O + 36\text{--}38\ ATP$$

glucose  oxygen  →  carbon dioxide  water

There is an uptake of oxygen during cellular respiration.

# 5.1 Fermentation

Yeasts (unicellular fungi) can use several types of sugars as an energy source. Glucose and fructose are monosaccharides; sucrose is a disaccharide that contains glucose and fructose. Fructose can be converted to glucose, the usual molecule acted on by yeast. In the following Experimental Procedure, by measuring the amount of carbon dioxide given off in a respirometer, you will test several of these food sources for their ability to ferment. A **respirometer** is a device for measuring the amount of gas given off and/or consumed.

## Experimental Procedure: Fermentation

### Respirometer Practice

1. Completely fill a small tube (15 × 125 mm) with water (Fig. 5.1).
2. Invert a large tube (20 × 150 mm) over the small tube, and with your finger or a pencil, push the small tube up into the large tube until the upper lip of the small tube is in contact with the bottom of the large tube.
3. Quickly invert both tubes. Do not permit the small tube to slip away from the bottom of the large tube. A little water will leak out of the small tube and be replaced by an air bubble.
4. Practice this inversion until the bubble in the small tube is as small as you can make it.

**Figure 5.1** **Respirometer for yeast experiment.**
Place a small tube inside a large tube. Hold the small tube in place as you rotate the entire apparatus, and an air bubble will form in the small tube.

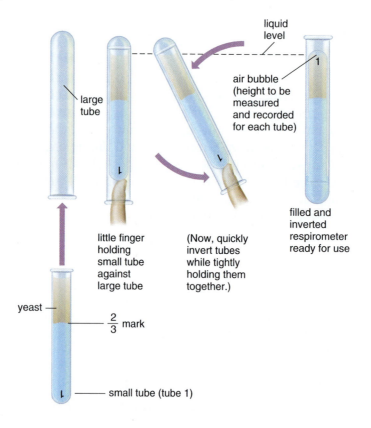

### Testing Food Samples

Have ready four large test tubes. With a wax pencil, label and mark off a small test tube at the 2/3-full level. Use this tube to mark off three other small tubes at the same level.

1.  Label and fill the small tubes as directed, and record the contents in Table 5.1.

    Tube 1   Fill to the mark with *glucose solution.*

    Tube 2   Fill to the mark with *fructose solution.*

    Tube 3   Fill to the mark with *sucrose solution.*

    Tube 4   Fill to the mark with *distilled water.*

2.  Resuspend a yeast solution each time and fill all four tubes to the top with yeast suspension (Fig. 5.1).
3.  Slide the large tubes over the small tubes and invert them in the way you practiced. This will mix the yeast and sugar solutions.
4.  Place the respirometers in a tube rack, and measure the initial height of the air space in the rounded bottom of the small tube. Record the height in Table 5.1.
5.  Place the respirometers in an incubator or in a warm water bath maintained at 37°C. Note the time, and allow the respirometers to incubate about 20 minutes (incubator) or one hour (water bath). However, watch your respirometers and if they appear to be filling with gas quite rapidly, stop the incubation when appropriate. During the incubation period, begin the Experimental Procedure on cellular respiration (see p. 50).
6.  At the end of the incubation period, measure the final height of the gas bubble, and record it in Table 5.1. Calculate the net change, and record it in Table 5.1.

| Table 5.1 | Fermentation by Yeast | | | | |
|---|---|---|---|---|---|
| **Tube** | **Contents** | **Initial Gas Height** | **Final Gas Height** | **Net Change** | **Conclusion** |
| 1 | | | | | |
| 2 | | | | | |
| 3 | | | | | |
| 4 | | | | | |

### Conclusions

*   From your results, conclude how the sugars tested compare as a food source for yeast fermentation. Enter your conclusions in Table 5.1.
*   How do you know that the yeast cells were respiring anaerobically (in the absence of oxygen) and not aerobically (in the presence of oxygen)?

    _____

*   Why did the gas bubbles increase in size? _____

*   Speculate on why sucrose is not as good a food source as fructose and glucose. _____

    _____

*   Which respirometer was the control? _____

## 5.2 Cellular Respiration

As indicated in the cellular respiration equation given in the introduction to this laboratory, oxygen gas is consumed during cellular respiration and carbon dioxide gas is given off. The uptake of oxygen is the evidence that an organism is carrying on cellular respiration. It is possible to use potassium hydroxide (KOH) to remove carbon dioxide as it is given off. The equation for this reaction is:

$$CO_2 + 2\ KOH \longrightarrow K_2CO_3 + H_2O$$
$$\text{solid}$$
$$\text{potassium}$$
$$\text{carbonate}$$

### Experimental Procedure: Cellular Respiration

1. Obtain a volumeter, an apparatus that measures changes in gas volumes. Remove the three vials from the volumeter. Remove the stoppers from the vials and the vials from the volumeter. Label the vials 1, 2, and 3.

> **Caution:** Potassium hydroxide (KOH) is a strong, caustic base. Exercise care in using this chemical, and follow your instructor's directions for disposal of these tubes. If any potassium hydroxide should spill on your skin, rinse immediately with water.

2. Using the same amounts, place a small wad of absorbent cotton in the bottom of each vial. Without getting the sides of the vials wet, use a dropper to saturate the cotton with 15% potassium hydroxide (KOH). Place a small wad of dry cotton on top of the KOH-soaked absorbent cotton (Fig. 5.2).
3. Count 25 germinating soybean seeds and add to vial 1. Count 25 dry soybean seeds and add to vial 2. Also, add beads to vial 2 so that the volume occupied in vial 2 is approximately the same as in vial 1. Add only beads to vial 3 to bring to approximately the same volume.
4. Each stopper should have a vent (rubber tube) with a clamp and a graduated side arm. Remove the clamps from the vents. Adjust the graduated side arm until only about 5 mm to 1 cm protrudes through the stopper (Fig. 5.3).
5. With a dropper, add a drop of *Brodie manometer fluid* (or water colored with vegetable dye and a small amount of detergent) to each side arm.

**Figure 5.2** **Vials.**
In this experiment, three vials are filled as noted.

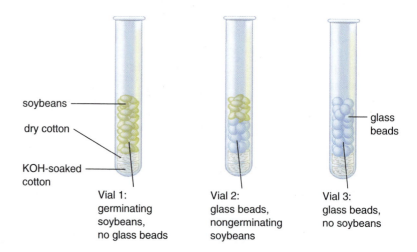

soybeans

dry cotton

KOH-soaked cotton

glass beads

Vial 1:
germinating
soybeans,
no glass beads

Vial 2:
glass beads,
nongerminating
soybeans

Vial 3:
glass beads,
no soybeans

6. Firmly place the stoppers in the vials. Adjust the stoppers until the side arms are parallel to the lab bench.

7. Adjust the location of the marker drop in the side arms with the assistance of a dry dropper in the vent. The marker drop of vial 3 (thermobarometer) should be in the middle of the side arm. The marker drop of vials 1 and 2 should be between 0.80 and 0.90 ml. Close the vent with the clamp when the marker drop is at the correct location.

8. Allow the respirometers to equilibrate for 5 minutes, and then record in Table 5.2, to the nearest 0.01 ml, the initial position of the marker drop in each graduated side arm.

9. Wait 10 minutes, and then record in Table 5.2 any change in the position of the marker drop. Wait 10 more minutes, and then record in Table 5.2 any change in the position of the marker drop. Then record in Table 5.2 the net change for each vial—that is, the initial reading for each vial minus the vial's 20-minute reading.

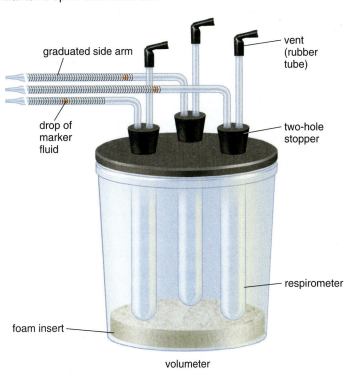

**Figure 5.3** **Volumeter containing three respirometers.** In this experiment, the respirometers are vials with graduated side arms attached. Oxygen uptake is measured by movement of a marker drop in each side arm.

10. Did the marker drop change in vial 3 (glass beads)? _____ By how much? _____ Enter this number in the "Correction" column of Table 5.2, and use this number to correct the net change you observed in vials 1 and 2. (This is a correction for any change in volume due to atmospheric pressure changes or temperature changes.) This will complete Table 5.2.

| Table 5.2 | Cellular Respiration | | | | | |
|---|---|---|---|---|---|---|
| Vial Contents | Initial Reading | Reading After 10 Minutes | Reading After 20 Minutes | Net Change | Correction | (Corrected) Net Change |
| 1 Germinating soybeans | | | | | | |
| 2 Nongerminating soybeans Glass beads | | | | | | |
| 3 Glass beads | | | | | | |

## Conclusions

- In which vial did the water recede? _____ State the vial contents. _____
  Is this the vial that carried on cellular respiration? _____
- Why did the water recede?_____

- Why was it necessary to absorb the carbon dioxide? _____ 5–6

_____

- In the soybean experiment, you were measuring the change in volume of what gas? _____

- Which respirometer in the soybean experiment was the control? _____

_____ 1. Both cellular respiration and fermentation begin with what molecule?

_____ 2. What reactant needed for cellular respiration is absent from the fermentation reaction?

_____ 3. What gas do organisms give off when they carry out cellular respiration?

_____ 4. Both fermentation and cellular respiration provide what molecule needed by cells?

_____ 5. Do plant cells or animal cells carry on cellular respiration?

_____ 6. Which process, fermentation or cellular respiration, results in a product that contains C—H bonds?

_____ 7. Name the device that can measure the amount of gas given off by yeast.

_____ 8. Yeast cells carry out fermentation when they are supplied with what type of molecule?

_____ 9. During the fermentation experiment, the gas bubble got larger. What gas was causing this increase?

_____ 10. What role was played by KOH in the soybean experiment?

_____ 11. In the soybean experiment, what do you call the tube that contains only glass beads?

_____ 12. What gas is being taken up when the marker in the side arm of a respirometer moves toward a tube that contains soybeans?

### Thought Questions

13. Why is it reasonable that, of the three sugars (glucose, fructose, and sucrose), glucose would result in the most activity during the fermentation experiment?

14. If you performed the cellular respiration experiment without soaking the cotton with KOH, what results would you predict? Why?

# 6
# Photosynthesis

## Introduction

The overall equation for **photosynthesis** is

$$CO_2 + H_2O \xrightarrow{\text{solar energy}} (CH_2O)_n + O_2$$

In this equation, $(CH_2O)$ represents any general carbohydrate. Sometimes, this equation is multiplied by 6 so that glucose $(C_6H_{12}O_6)$ appears as a product of photosynthesis. Photosynthesis takes place in chloroplasts (Fig. 6.1). Here membranous thylakoids are stacked in grana surrounded by the stroma. During the light reactions, pigments within the membranes, notably the chlorophylls, of thylakoids absorb solar energy, water $(H_2O)$ is split, and oxygen $(O_2)$ is released. The Calvin cycle reactions occur within the stroma. During these reactions, carbon dioxide $(CO_2)$ is reduced and solar energy is now stored in a carbohydrate $(CH_2O)$.

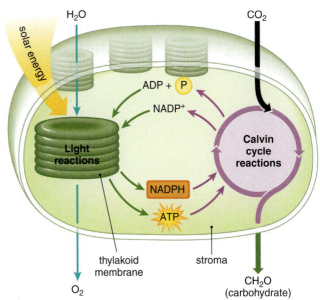

**Figure 6.1** Overview of photosynthesis.

# 6.1 Plant Pigments

The principal pigment in the thylakoids of plants is **chlorophyll *a*. Chlorophyll *b*, carotenes,** and **xanthophylls** play a secondary role by transferring the energy they absorb to chlorophyll *a* for use in photosynthesis.

**Chromatography** is a technique that separates molecules from each other on the basis of their solubility in particular solvents. The solvents used in the following Experimental Procedure are petroleum ether and acetone, which have no charged groups and are therefore nonpolar. As a nonpolar solvent moves up the chromatography paper, the pigment moves along with it. The more nonpolar a pigment, the more soluble it is in a nonpolar solvent and the faster and farther it proceeds up the chromatography paper.

### *Experimental Procedure: Plant Pigments*

> **Caution:** Ether (part of the chromatography solution) is toxic and extremely flammable. Do not breathe the fumes, and do not place the chromatography solution near any source of heat. A fume (ventilation) hood is recommended.

1. Assemble a chromatography apparatus (large, dry test tube and cork with a hook) and a strip of precut chromatography paper (handle by the top only) (Fig. 6.2). Attach the paper strip to the hook and test for fit. The paper should hang straight and barely touch the bottom of the test tube; trim if necessary. Measure 2 cm from the bottom of the paper and place a small dot with a pencil (not a pen). With a wax pencil, mark the test tube 1 cm below where the dot is with the stopper in place. Set the apparatus in a test-tube rack.

### Figure 6.2  Paper chromatography.
The paper must be cut to size and arranged to hang down without touching the sides of a dry tube. Then the pigment (chlorophyll) solution is applied to a designated spot. The chromatogram develops after the spotted paper is suspended in the chromatography solution.

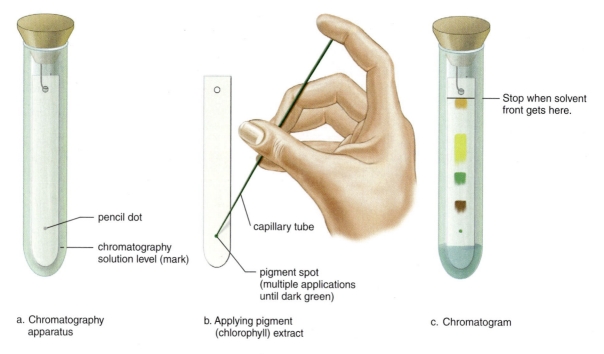

pencil dot

chromatography solution level (mark)

a. Chromatography apparatus

capillary tube

pigment spot (multiple applications until dark green)

b. Applying pigment (chlorophyll) extract

Stop when solvent front gets here.

c. Chromatogram

2. Prepare (or obtain) a plant *pigment extract,* as directed by your instructor.
3. Place the premarked chromatography paper strip onto a paper towel.
4. Fill a capillary tube by placing it into the extract. (It will fill by its own capillary action.)
5. Repeatedly apply the *pigment extract* to the pencil dot on the chromatographic strip. Let the spot dry between each application. Try to obtain a small dark green spot. (Placing your index finger over the end of the capillary tube will help keep the dot small.)
6. In a **fume hood,** add *chromatography solution* to the mark you made earlier. Do not submerge the pigment spot. Set the apparatus in a test-tube rack and close the chromatography apparatus tightly. Do not shake the test tube during the chromatography.
7. Allow approximately 10 minutes for your chromatogram to develop, but check it frequently so that the pigments do not reach the top of the paper.
8. When the solvent front has moved to within 1 cm of the upper edge of the paper (Fig. 6.2c), remove your chromatogram. Close the apparatus tightly. With a pencil, lightly mark the location of the solvent front, and allow the chromatogram to dry in the fume hood.
9. Identify the pigment bands. Beta-carotenes are represented by the bright orange-yellow band at the top. Xanthophylls are yellow and may be represented in multiple bands. The blue-green band is chlorophyll *a,* and the lowest, olive-green band is chlorophyll *b.* Which pigment is the most nonpolar (that is, has the greatest affinity for the nonpolar solvent)? _____
10. Calculate the $R_f$ (ratio-factor) values for each pigment. For these calculations, mark the center of the initial pigment spot. This will be the starting point for all measurements. Also mark the midpoints of each pigment and the solvent front. Measure the distance between points for each pigment in millimeters (mm), and record these values in Table 6.1. Then use the following formula, and enter your $R_f$ values in Table 6.1:

$$R_f = \frac{\text{distance moved by pigment}}{\text{distance moved by solvent}}$$

| Table 6.1 | $R_f$ (Ratio-Factor) Values for Each Pigment | |
|---|---|---|
| Pigments | Distance Moved (mm) | $R_f$ Values |
| Beta-carotenes | | |
| Xanthophylls | | |
| Chlorophyll *a* | | |
| Chlorophyll *b* | | |
| Solvent | | |

11. Do your results suggest that the chemical characteristics of these pigments might differ?_____
    Why? _____

# 6.2 Solar Energy

During photosynthesis, solar energy is transformed into the chemical energy of a carbohydrate $(CH_2O)_n$. Without solar energy, photosynthesis would be impossible.

Release of oxygen from a plant indicates that photosynthesis is occurring. *Verify that photosynthesis releases oxygen by writing the equation for photosynthesis below.*

## Role of White Light

White (sun) light contains different colors of light, as is demonstrated when white light passes through a prism (Fig. 6.3). White light is the best for photosynthesis because it contains all the colors of light. The oxygen released from photosynthesis is taken up by a plant when cellular respiration occurs. This must be taken into account when the rate of photosynthesis is calculated.

**Figure 6.3   White light.**
White light is made up of various colors, as can be seen when white light passes through a prism.

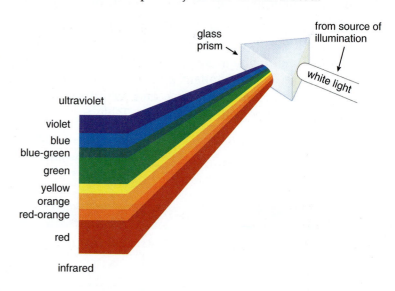

ultraviolet
violet
blue
blue-green
green
yellow
orange
red-orange
red
infrared

glass prism
from source of illumination
white light

## *Experimental Procedure: White Light*

1. Place a generous quantity of *Elodea* with the cut end up (make sure the cuts are fresh) in a test tube with a rubber stopper containing a bent piece of glass tubing, as illustrated in Figure 6.4. When assembled, this is your volumeter for studying the need for light in photosynthesis. (Do not hold the volumeter in your hand, as body heat will also drive the reaction forward). Your instructor will show you how to fix the volumeter in an upright position.

2. Before stoppering the test tube, add sufficient 3% *sodium bicarbonate* $(NaHCO_3)$ solution so that when the rubber stopper is inserted into the tube, the solution comes to rest at about 1/4 the length of the bent glass tubing. Mark this location on the glass tubing with a wax pencil.

3. Place a beaker of plain water next to the *Elodea* tube to serve as a heat absorber. Place a lamp (150 watt) next to the beaker. The tube, beaker, and lamp should be as close to one another as possible.

4. Turn on the lamp. As soon as the edge of the solution in the tubing begins to move, time the reaction for 10 minutes. Be careful not to bump the tubing or to readjust the stopper, or your readings will be altered. After 10 minutes, mark the edge of the solution, and measure in

   millimeters the distance the edge moved: _____ mm/10 min. This is **net photosynthesis,** a measurement that does not take into account the oxygen used up for cellular respiration.

   Record your results in Table 6.2. Why did the edge move forward?

## Figure 6.4 Volumeter.
A volumeter apparatus is used to study the role of light in photosynthesis.

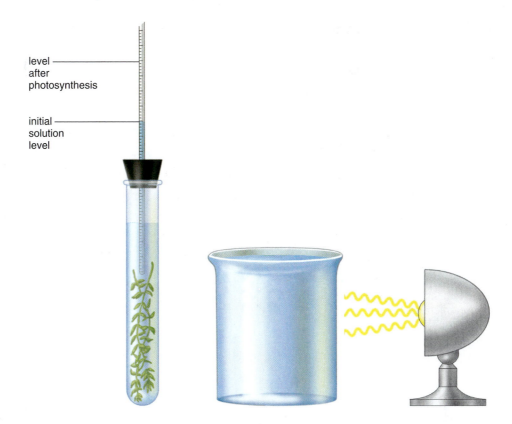

level after photosynthesis

initial solution level

5. Carefully wrap the tube containing *Elodea* in aluminum foil, and record here the length of time it takes for the edge of the solution in the tubing to recede 1 mm: _____. Convert your measurement to _____ mm/10 min., and record this value for **cellular respiration** in Table 6.2. (Do not use a minus sign, even though the edge receded.) Why does cellular respiration, which occurs in a plant all the time, cause the edge to recede? _____

_____

6. If the *Elodea* had not been respiring in step 4, how far would the edge have moved? _____ mm/10 min. This is **gross photosynthesis** (net photosynthesis + cellular respiration). Record this number in Table 6.2.

7. Calculate the **rate of photosynthesis** (mm/hr) by multiplying gross photosynthesis (mm/10 min) by 6 (that is, 10 min × 6 = 60 min = 1 hr): _____ mm/hr. Record this value in Table 6.2.

| Table 6.2 | Rate of Photosynthesis (White Light) |
|---|---|
| | Data |
| Net photosynthesis (white light) | |
| Cellular respiration (no light) | |
| **Gross photosynthesis** (net + cellular respiration) | (mm/10 min) |
| **Rate of photosynthesis** | (mm/hr) |

## Role of Green Light

Green light is only one part of white light (see Fig. 6.3). Plant pigments absorb certain colors of light better than other colors (Fig. 6.5). According to Figure 6.5, what color light do the chlorophylls absorb best? _____ Least? _____

What color light do the carotenoids (carotenes and xanthophylls) absorb best? _____

Least? _____

Does photosynthesis use green light? _____

The following Experimental Procedure will test your answer.

**Figure 6.5** Action spectrum for photosynthesis.

The action spectrum for photosynthesis is the sum of the absorption spectrums for the pigments chlorophyll a, chlorophyll b, and carotenoids.

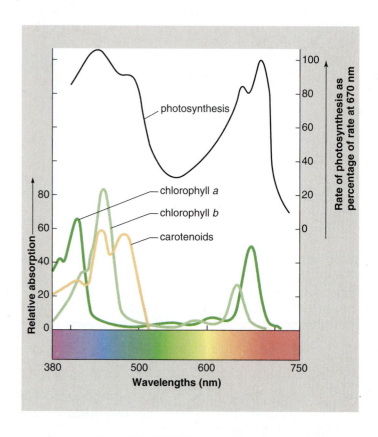

## Experimental Procedure: Green Light

1.  Add three drops of green dye (or use a green cellophane wrapper) to the beaker of water used in the previous Experimental Procedure until there is a distinctive green color. Remove all previous wax pencil marks from the glass tubing.
2.  Record in Table 6.3 your data for gross photosynthesis (mm/10 min) and for rate of photosynthesis for white light (mm/hr) from Table 6.2.
3.  Turn on the lamp. Mark the location of the edge of the solution on the glass tubing. As soon as the edge begins to move, time the reaction for 10 minutes. After 10 minutes, mark the edge of the solution, and measure in millimeters the distance the edge moved. Net photosynthesis

    for green light = _____ mm/10 min.
4.  Carefully wrap the tube containing *Elodea* in aluminum foil, and record here the length of time

    it takes for the edge of the solution in the tubing to recede 1 mm: _____. Convert your

    measurement to _____ mm/10 min.

5. Calculate gross photosynthesis for green light (mm/10 min) as you did for white light and record your data in Table 6.3.

6. Calculate rate of photosynthesis for green light (mm/hr) as you did for white light and record your data in Table 6.3.

7. Average and record the Table 6.3 class data for both white light and green light and record these averages in Table 6.3.

8. The following equation shows the rate of photosynthesis (green light) as a percentage of the rate of photosynthesis (white light):

$$\text{percentage} = \frac{\text{rate of photosynthesis (green light)}}{\text{rate of photosynthesis (white light)}} \times 100$$

This percentage, based on your data in Table 6.3 = _____. This percentage, based on class data in Table 6.3 = _____. Record these values in Table 6.3.

| Table 6.3 | Rate of Photosynthesis (Green Light) | |
|---|---|---|
| | Your Data | Class Data |
| **Gross photosynthesis (mm/10 min)** | | |
| White (from Table 6.2) | | |
| Green | | |
| **Rate of photosynthesis (mm/hr)** | | |
| White (from Table 6.2) | | |
| Green | | |

## Conclusions

- Explain why the rate of photosynthesis with green light is only a portion of the rate of photosynthesis with white light. _____
_____

- How does the percentage based on your data differ from that based on class data?
_____
_____

# 6.3 Carbon Dioxide Uptake

During the second stage of photosynthesis, the plant takes up carbon dioxide ($CO_2$) and reduces it to a carbohydrate, such as glucose ($C_6H_{12}O_6$). Therefore, the carbon dioxide in the solution surrounding *Elodea* should disappear as photosynthesis takes place.

## Experimental Procedure: Carbon Dioxide Uptake

1. Temporarily remove the *Elodea* from the test tube. Empty the sodium bicarbonate ($NaHCO_3$) solution from the test tube, rinse the test tube thoroughly, and fill with a phenol red solution diluted to a faint pink. (Add more water if the solution is too dark.) Phenol red is a pH indicator that turns yellow in an acid and red in a base.

2. Blow *lightly* on the surface of the solution. Stop blowing as soon as the surface color changes to yellow. Then shake the test tube until the rest of the solution turns yellow.

   Blowing onto the solution adds what gas to the test tube? _____ When carbon dioxide combines with water, it forms carbonic acid. What causes the color change?

   _____

3. Thoroughly rinse the *Elodea* with distilled water, return it to the test tube with the phenol red solution, and assemble your volumeter as before.

4. If you used green dye, change the water in the beaker to remove the green solution.

5. Turn on the lamp and wait until the edge of the solution just begins to move. Note the time. Observe until you note a change in color. Record your results in the appropriate column of Table 6.4.

6. Hypothesize why the solution in the test tube eventually turned red. _____

   _____

## Use of a Control

Scientists are more confident of their results when an experimental procedure includes a control. Controls undergo all the steps in the experiment except the one being tested.

- Considering the test sample in Table 6.4, suggest a possible control sample for this experiment:

  _____

- Ask your instructor if you can perform this procedure. Both the control and test sample should be done at the same time.

- Record your results in Table 6.4. Why should all experiments have a control? _____

  _____

| Table 6.4 | Carbon Dioxide Uptake |
| --- | --- |
| Tube | Time for Color Change |
| Test sample: *Elodea* + phenol red solution + $CO_2$ | |
| Control sample: | |

## 6.4 Carbon Cycle

In this laboratory, you have demonstrated a relationship between cellular respiration and photosynthesis. Animals produce the carbon dioxide used by plants to carry out photosynthesis. Plants produce the food and oxygen that they and animals require to carry out cellular respiration. This relationship is illustrated in Figure 6.6 and can be represented by the following equation:

$$C_6H_{12}O_6 + 6\ O_2 \underset{\text{photosynthesis}}{\overset{\substack{\text{cellular}\\\text{respiration}}}{\rightleftharpoons}} 6\ CO_2 + 6\ H_2O + \text{energy}$$

1. Which organelle in plants carries out the reaction in the previous equation in the reverse (right-to-left) direction? _____

2. Pertaining to photosynthesis, the energy in the equation is provided by _____.

3. Which organelle in plants and animals is involved in carrying out the reaction in this equation in the forward direction? _____

4. Pertaining to cellular respiration, the energy in the equation becomes chemical bond energy in what molecule? _____

5. Would it be correct to say that solar energy eventually becomes the chemical bond energy in ATP? _____ Why? _____

6. Considering that both plants and animals carry on cellular respiration, revise Figure 6.6 to improve its accuracy.

**Figure 6.6** **Photosynthesis and cellular respiration.**
Animals are dependent on plants for a supply of oxygen, and plants are dependent on animals for a supply of carbon dioxide.

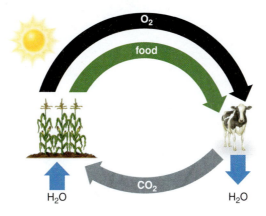

1. Where do the light reactions of photosynthesis take place?
2. What procedure did you use to separate plant pigments?
3. What determines the distance a pigment moves up the paper?
4. Where do plants ordinarily get the energy they need to carry on photosynthesis?
5. Green plants do not absorb what color of light?
6. Blue, red, and green light are all present in what color of light?
7. Why do blue and red light, but not green, promote photosynthesis?
8. Does *Elodea* respire in the light or in the dark?
9. If net photosynthesis is 5 mm/10 min and cellular respiration is .5 mm/10 min, how much is gross photosynthesis?
10. Phenol red turns what color when carbon dioxide is added?
11. What happens to carbon dioxide during photosynthesis?
12. What two substances do plants provide for us?

## Thought Questions

13. Some plants are colorless. Do you predict that they carry on photosynthesis? Explain.

14. Suppose there were single-celled protozoans (nonphotosynthetic) in the test tube with *Elodea* when you did the white light experiment. Could you still calculate *Elodea*'s rate of photosynthesis? Explain.

# 7

# Mitosis and Meiosis

## Learning Objectives

**7.1 The Cell Cycle**
- What are the four stages of the cell cycle?
- Name and describe the phases of mitosis, with attention to the movement of chromosomes.
- Identify the phases of mitosis in models and microscopic slides.
- Contrast the associated structures of animal and plant cell mitosis.
- Describe animal and plant cell cytokinesis.

**7.2 Meiosis**
- Name and describe the phases of meiosis I, with attention to the movement of chromosomes.
- Name and describe the phases of meiosis II, with attention to the movement of chromosomes.

**7.3 Mitosis Versus Meiosis**
- Contrast the behavior of chromosomes during mitosis with the behavior of chromosomes during meiosis I.
- Contrast the behavior of chromosomes during mitosis and meiosis II.

## Introduction

Dividing cells experience nuclear division, cytoplasmic division, and a period between divisions called interphase. During **interphase,** the nucleus appears normal, and the cell is performing its usual cellular functions. Also, the cell is increasing all of its components, including such organelles as the mitochondria, ribosomes, and centrioles, if present. DNA replication (making an exact copy of the DNA) occurs toward the end of interphase. Thereafter, the chromosomes, which contain DNA, are duplicated and contain two chromatids held together at a **centromere.** These chromatids are called **sister chromatids.**

During nuclear division, called **mitosis,** the new nuclei receive the same number of chromosomes as the parental nucleus. When the cytoplasm divides, a process called **cytokinesis,** two daughter cells are produced. In multicellular organisms, mitosis permits growth and repair of tissues. In eukaryotic, unicellular organisms, mitosis is a form of asexual reproduction. Sexually reproducing organisms use another form of nuclear division, called **meiosis.** In animals, meiosis is a part of gametogenesis, the production of gametes (sex cells) and occurs in organ caused gonads. The gametes are sperm in male animals and eggs in female animals. As a result of meiosis, the daughter cells have half the number of chromosomes as the parental cell. Because crossing-over of genetic material takes place and the chromosomes occur in various combinations in the daughter cells, meiosis contributes to recombination of genetic material and to variation among sexually reproducing organisms.

This laboratory examines both mitotic and meiotic cell division to show their similarities and differences. At the start of both types of divisions, the parental nucleus, surrounded by a double membrane (the nuclear envelope), contains one or more **nucleoli** (concentrated regions of RNA) and **chromatin** (threadlike strands of DNA) suspended in a transparent liquid called **nucleoplasm.** During division, chromatin condenses so that the chromosomes are visible, the nuclear envelope fragments, and a spindle appears. Spindle fibers assist the movement of chromosomes, which occurs at this time.

# 7.1 The Cell Cycle

As stated in the "Introduction," the period between cell divisions is known as interphase. Because early investigators noted little visible activity between cell divisions, they dismissed this period as a resting state. But when they discovered that DNA replication and chromosome duplication occur during interphase, the **cell cycle** concept was proposed. Investigators have also discovered that cytoplasmic organelle duplication occurs during interphase, as does synthesis of the proteins involved in regulating cell division. Thus, the cell cycle can be broken down into four stages (Fig. 7.1). State the event of each stage on the line provided:

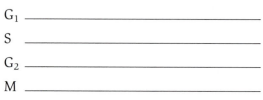

$G_1$ _____

S  _____

$G_2$ _____

M  _____

The time required for the entire cell cycle varies according to the organism, but 18–24 hours is typical for animal cells. Mitosis (including cytokinesis, if it occurs) lasts less than an hour to slightly more than 2 hours; for the rest of the time, the cell is in interphase.

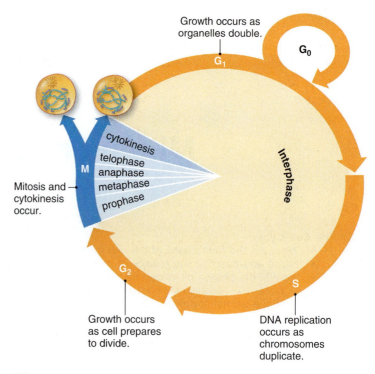

Growth occurs as organelles double.

$G_0$

$G_1$

Interphase

cytokinesis
telophase
anaphase
metaphase
prophase

M

Mitosis and cytokinesis occur.

$G_2$

Growth occurs as cell prepares to divide.

S

DNA replication occurs as chromosomes duplicate.

**Figure 7.1  The cell cycle.**
Cells go through a cycle that consists of four stages: $G_1$, S, $G_2$, and M. The major activity for each stage is given. Some cells can exit $G_1$ and enter a $G_0$ stage.

| Table 7.1 | Structures Associated with Mitosis |
|---|---|
| **Structure** | **Description** |
| Nucleus | A large organelle containing the chromosomes and acting as a control center for the cells |
| Chromosome | Rod-shaped body in the nucleus seen during mitosis and meiosis and that contains DNA and, therefore the hereditary units, or genes |
| Nucleolus | An organelle found inside the nucleus; composed largely of RNA for ribosome formation |
| Spindle | Microtubule structure that brings about chromosome movement during cell division |
| Chromatids | The two identical parts of a chromosome following DNA replication |
| Centromere | A constriction where the sister chromatids of a chromosome are held together |
| Centrosome | The central microtubule-organizing center of cells; consists of granular material; in animal cells, contains two centrioles |
| Centriole* | A short, cylindrical organelle in animal cells that contains microtubules and is associated with the formation of the spindle during cell division |
| Aster* | Short, radiating fibers produced by the centrioles; important during mitosis and meiosis |

*Animal cells only

## Mitosis

**Mitosis** is nuclear division that results in two new nuclei, each having the same number of chromosomes as the original nucleus. The parental cell is the cell that divides, and the resulting cells are called daughter cells.

When cell division is about to begin, chromatin starts to condense and compact to form visible, rodlike sister chromatids held together at the **centromere** (Fig. 7.2a). Label the sister chromatids and the centromere in the drawing of a duplicated chromosome in Figure 7.2b. This illustration represents a chromosome as it would appear just before nuclear division occurs.

### Figure 7.2  Duplicated chromosomes.
DNA replication results in a duplicated chromosome that consists of two sister chromatids held together at a centromere. *a.* Scanning electron micrograph of a duplicated chromosome. *b.* In this drawing, one chromatid is indicated.

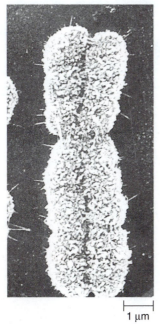

a. Photomicrograph of highly compacted chromosome

1 μm

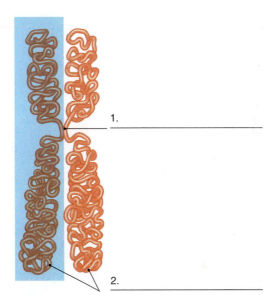

1. _____

2. _____

b. Diagram of less compacted chromosome

## Mitotic Spindle

Table 7.1 lists the structures that play a role during mitosis. The spindle is a structure that appears and brings about an orderly distribution of chromosomes to the daughter cell nuclei. A spindle has fibers that stretch between two poles (ends). Spindle fibers are bundles of microtubules, protein cylinders found in the cytoplasm that can assemble and disassemble. The **centrosome,** the main microtubule-organizing center of the cell, divides before mitosis so that during mitosis each pole of the spindle has a centrosome. Animal cells contain two barrel-shaped organelles called centrioles in each centrosome and asters, arrays of short microtubules radiating from the poles (see Fig. 7.3). That plant cells lack centrioles suggests that centrioles are not required for spindle formation.

## Mitosis Phases

Mitosis is the type of nuclear division that (1) occurs in the body (somatic) cells; (2) results in two daughter cells because there is only one round of division; and (3) keeps the chromosome number constant (same as the parent cell). Before mitosis begins, each chromosome is already duplicated and composed of two sister chromatids held together by a centromere (see Fig. 7.2). Sister chromatids are

**Figure 7.3** **Phases of mitosis in animal cells.**

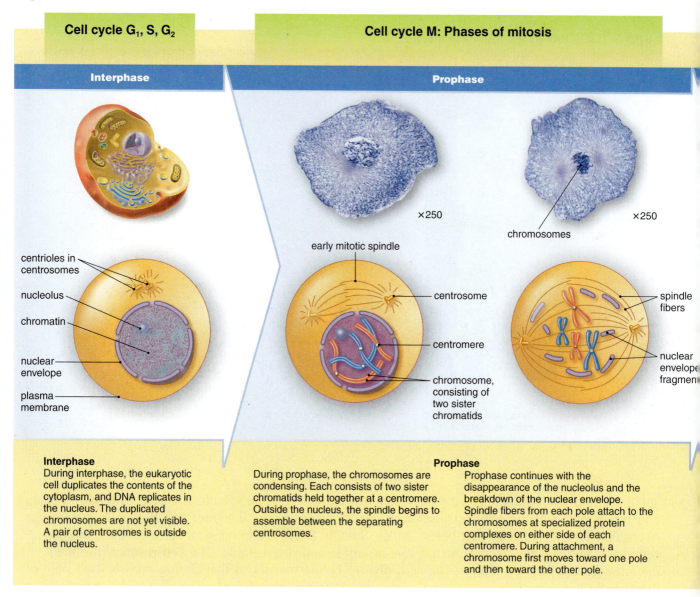

**Interphase**
During interphase, the eukaryotic cell duplicates the contents of the cytoplasm, and DNA replicates in the nucleus. The duplicated chromosomes are not yet visible. A pair of centrosomes is outside the nucleus.

**Prophase**
During prophase, the chromosomes are condensing. Each consists of two sister chromatids held together at a centromere. Outside the nucleus, the spindle begins to assemble between the separating centrosomes.

**Prophase**
Prophase continues with the disappearance of the nucleolus and the breakdown of the nuclear envelope. Spindle fibers from each pole attach to the chromosomes at specialized protein complexes on either side of each centromere. During attachment, a chromosome first moves toward one pole and then toward the other pole.

genetically identical—that is, they contain exactly the same alleles. During mitosis, the duplicated chromosomes become attached to the mitotic spindle by their centromeres.

When the centromeres divide, the spindle distributes daughter chromosomes equally to its poles (opposite ends). Therefore, each daughter cell gets a complete copy and one of each kind of chromosome. Counting the number of centromeres tells you the number of chromosomes. The phases of mitosis in animal cells is shown in Figure 7.3.

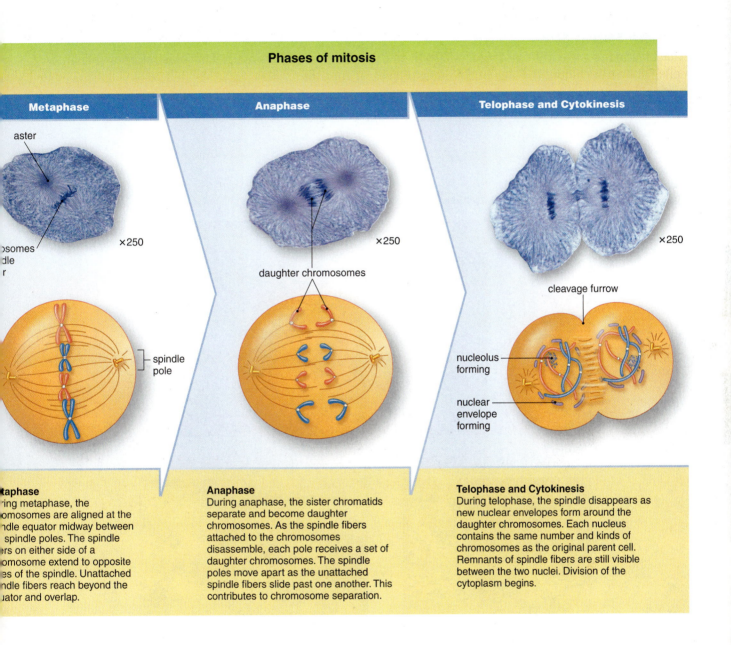

**Phases of mitosis**

**Metaphase**

aster

×250

osomes
dle
r

spindle
pole

**Anaphase**

×250

daughter chromosomes

**Telophase and Cytokinesis**

×250

cleavage furrow

nucleolus
forming

nuclear
envelope
forming

**taphase**
ring metaphase, the
omosomes are aligned at the
ndle equator midway between
spindle poles. The spindle
ers on either side of a
omosome extend to opposite
es of the spindle. Unattached
ndle fibers reach beyond the
uator and overlap.

**Anaphase**
During anaphase, the sister chromatids separate and become daughter chromosomes. As the spindle fibers attached to the chromosomes disassemble, each pole receives a set of daughter chromosomes. The spindle poles move apart as the unattached spindle fibers slide past one another. This contributes to chromosome separation.

**Telophase and Cytokinesis**
During telophase, the spindle disappears as new nuclear envelopes form around the daughter chromosomes. Each nucleus contains the same number and kinds of chromosomes as the original parent cell. Remnants of spindle fibers are still visible between the two nuclei. Division of the cytoplasm begins.

### Animal Mitosis Models

1. Using the previous descriptions and Figure 7.3 as a guide, identify the phases of animal cell mitosis in models of animal cell mitosis.
2. Each species has its own chromosome number. Counting the number of centromeres tells you the number of chromosomes in the models. What is the number of chromosomes observed in

   each nucleus of the cells? _____

### Whitefish Blastula Slide

The blastula is an early embryonic stage in the development of animals. The **blastomeres** (blastula cells) shown in Figure 7.4 are in different phases of mitosis.

1. Examine a prepared slide of whitefish blastula cells undergoing mitotic cell division.
2. Try to find a cell in each phase of mitosis. Have a partner or your instructor check your identification.

## Figure 7.4 Whitefish blastula.
The blastula is an embryonic stage in which the cells are undergoing mitosis. (*a–d:* Magnification ×250)

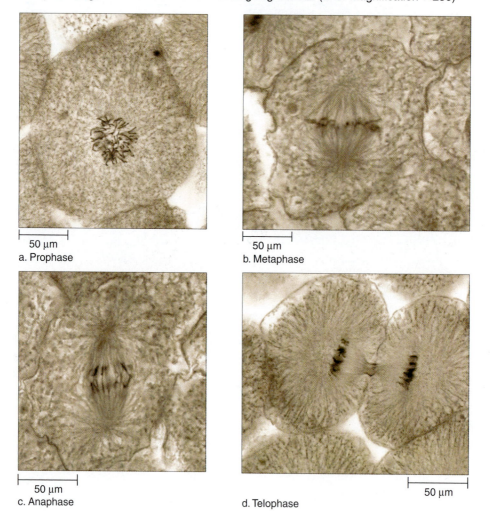

a. Prophase

b. Metaphase

c. Anaphase

d. Telophase

## Plant Mitosis Phases

The phases of plant mitosis are exactly the same as those of animal mitosis (Fig. 7.5). During early prophase, the chromatin condenses into scattered, previously duplicated chromosomes, and the spindle forms; during late prophase, chromosomes attach to spindle fibers; during metaphase, the chromosomes are at the metaphase plate of the spindle; during anaphase, the daughter chromosomes move to the poles of the spindle; and during telophase, division of the cytoplasm by formation of a **cell plate** begins. *Write the name of the phase beneath each photo in Figure 7.5.*

Plant cells do not have centrioles and asters. However, plant cells do have centrosomes, the central microtubule-organizing center of the cell, and this accounts for the formation of a spindle.

### Figure 7.5   Mitosis in plant cells.

The chromosomes are stained blue, and microtubules of the spindle are stained red in these photos of a dividing African blood lily.

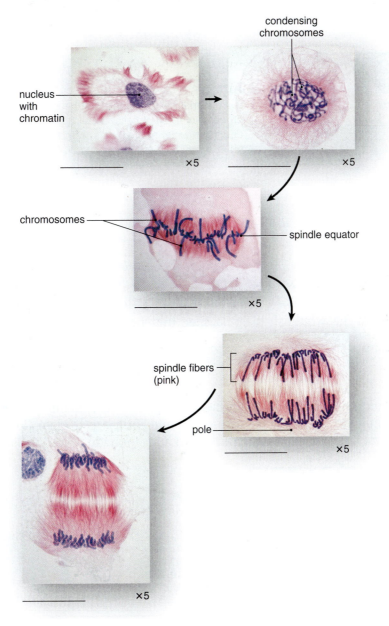

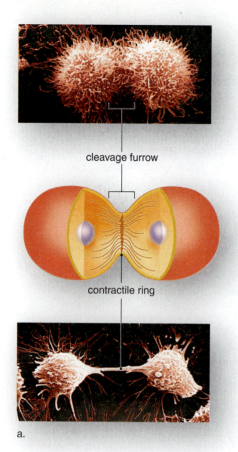

cleavage furrow

contractile ring

a.

daughter cells

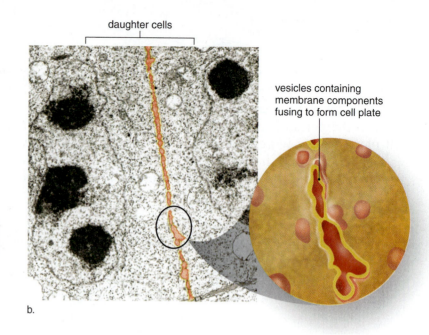

vesicles containing
membrane components
fusing to form cell plate

b.

**Figure 7.6** Cytokinesis.
a. In animal cells, a single cell becomes two cells by a furrowing
process. b. In plant cells, a cell plate forms midway between two
daughter nuclei and extends to the plasma membrane.
(a) Copyright by R.G. Kessel and C.Y. Shih, *Scanning Electron Microscopy in
Biology: A Student's Atlas on Biological Organization,* Springer-Verlag, 1974.

## Cytokinesis in Animal Cells

In animal cells, a **cleavage furrow,** an indentation of the membrane between the daughter nuclei,
begins as anaphase draws to a close (Fig. 7.6*a*). The cleavage furrow deepens as a band of actin
filaments called the contractile ring slowly constricts the cell, forming two daughter cells.

Were any of the cells of the whitefish blastula slide undergoing cytokinesis?

How do you know? _____

## Cytokinesis in Plant Cells

After mitosis, the cytoplasm divides by cytokinesis. In plant cells, membrane vesicles derived from
the Golgi apparatus migrate to the center of the cell and form a cell plate (Fig. 7.6*b*), the location
of a new plasma membrane for each daughter cell. Later, individual cell walls appear in this area.

## Summary of Mitotic Cell Division

1. Mitosis and cytokinesis results in _____ daughter cells.
2. The nuclei in the daughter cells have the _____ number of chromosomes as the parental
   cell had.
3. Mitosis is cell division in which the chromosome number _____.
4. Mitosis occurs during _____ and _____ of somatic tissues.

# 7.2 Meiosis

**Meiosis** is a form of nuclear division in which the chromosome number is reduced by half. While the nucleus of the parental cell has the diploid number of chromosomes, the daughter nuclei, after meiosis is complete, have the haploid number of chromosomes. In sexually reproducing species, meiosis must occur or the chromosome number would double with each generation.

A diploid nucleus contains **homologues,** also called homologous chromosomes. Homologues look alike and carry the genes for the same traits. Before meiosis begins, the chromosomes are already double stranded—that is, they contain sister chromatids.

## Experimental Procedure: Meiosis

In this exercise, you will use pop beads to construct homologues and move the chromosomes to simulate meiosis.

### Building Chromosomes to Simulate Meiosis

1.  Obtain the following materials: 48 pop beads of one color (e.g., red) and 48 pop beads of another color (e.g., blue) for a total of 96 beads in all; eight magnetic centromeres; and four centriole groups.
2.  Build a homologous pair of duplicated chromosomes using Figure 7.7a as a guide. Each chromatid will have 16 beads. Be sure to bring the centromeres of two units of the same color together so that they attract and link to form one duplicated chromosome. (One member of the pair will be red, and the other will be blue.)
3.  Build another homologous pair of duplicated chromosomes using Figure 7.7b as a guide. Each chromatid will have eight beads. Be sure to bring the centromeres of two units of the same color together so that they attract. (One member of the pair will be red, and the other will be blue.)
4.  Your chromosomes are the same as those in Figure 7.8. The red chromosomes were inherited from one parent, and the blue chromosomes were inherited from the other parent.

**Figure 7.7** **Two homologues of duplicated chromosomes.**

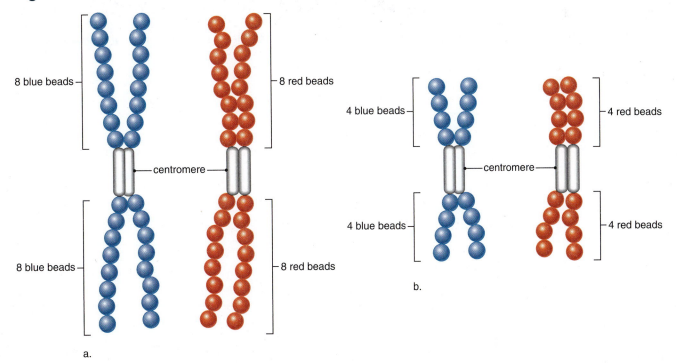

### Meiosis I

Meiosis requires two nuclear divisions and the first round during meiosis is called **meiosis I** (see Figure 7.8).

During prophase of meiosis I, the spindle appears while the nuclear envelope and nucleolus disappear. Homologues line up next to one another during a process called synapsis. During **crossing-over,** the nonsister chromatids of a homologue pair exchange genetic material. At metaphase I, the homologue pairs line up at the metaphase plate of the spindle. During anaphase I, homologues separate and the chromosomes (still composed of two chromatids) move to each pole. In telophase I, the nuclear envelope and the nucleolus reappear as the spindle disappears. Each new nucleus contains one from each pair of chromosomes.

### Prophase I

5. Using Figure 7.8 as a guide, put all four of the chromosomes you built in the center of your work area, which represents the nucleus. Place your two pairs of centrioles outside the nucleus.
6. Separate the pairs of centrioles, and move one pair to opposite poles of the nucleus.
7. Synapsis is the pairing of homologues during prophase I. Simulate synapsis by bringing the homologues together.
8. **Crossing-over** is a way to achieve genetic recombination during meiosis. Simulate crossing-over by exchanging the exact segments of two nonsister chromatids of a single homologous pair. Why

   use nonsister chromatids and not sister chromatids? _____

### Metaphase I

Position the homologues at the metaphase plate in such a way that the homologues are prepared to move apart toward the centrioles.

### Anaphase I

Separate the homologues, and move each one toward the opposite pole.

### Telophase I

9. During telophase I, the chromosomes are at the poles. What combinations of chromosomes are at the poles? Fill in the following blanks with the words *red-long, red-short, blue-long,* and *blue-short:*

   Pole A: _____ and _____

   Pole B: _____ and _____
10. What other combinations would have been possible? (*Hint:* Alternate the colors at metaphase I.)

    Pole A: _____ and _____

    Pole B: _____ and _____

### Conclusions

- Do the chromosomes inherited from the mother or father have to remain together following meiosis I? _____
- Name two ways that meiosis contributes to genetic variation:

    a. _____

    b. _____

### Interkinesis

**Interkinesis** is the period between meiosis I and meiosis II. In some species, daughter cells do not form, and meiosis II follows right after meiosis I. *Does DNA replication occur during interkinesis?* _____ Explain. _____

### Meiosis II

The second round of nuclear division during meiosis is called **meiosis II** (Figure 7.9).

During prophase of meiosis II, a spindle appears. Each chromosome attaches to the spindle independently. During metaphase II, the chromosomes are lined up at the metaphase plate. During anaphase II, the centromeres divide and the chromatids separate, becoming daughter chromosomes that move toward the poles. In telophase II, the spindle disappears as the nuclear envelope reappears. Meiosis II is exactly like mitosis except that the nuclei of the parental cell and the daughter cells are haploid.

#### Prophase II

1. Using Figure 7.9 as a guide, choose the chromosomes from one pole to represent those in the new parental cell undergoing meiosis II.
2. Place two pairs of centrioles at opposite sides of these chromosomes to form the new spindle.

#### Metaphase II

Move the duplicated chromosomes to the metaphase II metaphase plate. *How many chromosomes are at the metaphase II metaphase plate?* _____

#### Anaphase II

Pull the two magnets of each duplicated chromosome apart. *What does this action represent?* _____
_____

#### Telophase II

Put the chromosomes—each having one chromatid—at the poles (the new centrioles).

### Conclusions

- You worked with only one daughter cell from meiosis I as the new parental cell. Suppose you had worked with both daughter cells. How many cells would have been present when meiosis II was complete? _____
- How many chromosomes are in the parental cell undergoing meiosis II? _____
- How many chromosomes are in the daughter cell? _____ Explain. _____
_____

### Summary of Meiotic Cell Division

1. The parental cell has the diploid (2n) number of chromosomes, and the daughter cells have the _____ (n) number of chromosomes.
2. Meiosis is cell division in which the chromosome number _____.
   Whereas meiosis reduces the chromosome number, fertilization restores the chromosome number.
3. A zygote contains the same number of chromosomes as the parent, but are these exactly the same chromosomes? _____
4. What is another way that sexual reproduction results in genetic variation?
_____

## Meiosis Phases

Meiosis is the type of nuclear division that (1) occurs in the sex organs (testes and ovaries); (2) results in four cells because there are two rounds of cell division; and (3) reduces the chromosome number to half that of the parental cell. The phases of meiosis I are shown in Figure 7.8.

As discussed on page 72, crossing-over is a way to achieve genetic recombination during mitosis. Crossing-over temporarily holds the homologues together so that they interact with the spindle as if they were one chromosome. Eventually, the homologues separate, and this ensures that one duplicated chromosome of each pair reaches the daughter nuclei following meiosis I. The separation process has no restrictions: either homologue of a homologous pair may be passed on to a daughter

**Figure 7.8** **Meiosis I.**

As a result of meiosis I, the daughter cells are haploid and are genetically dissimilar. Each will contain a different combination of chromosomes and a different combination of alleles on the chromosomes. Note that when the homologues undergo synapsis during Prophase I, four chromatids are in a row. These four chromatids are called a tetrad. Also, note that when the homologues separate in Anaphase I, each chromosome is a dyad because it contains two chromatids.

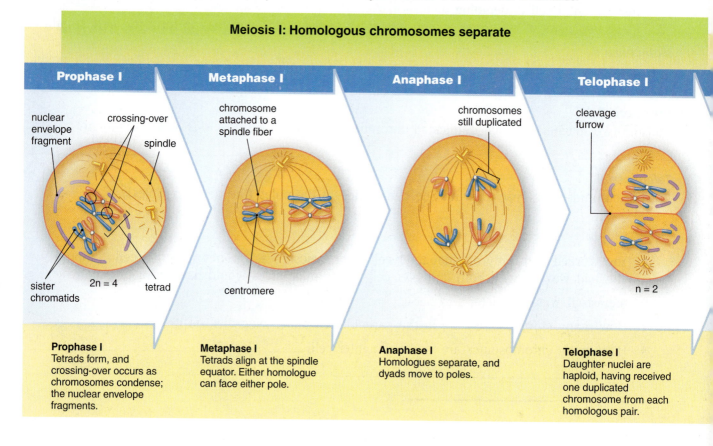

**Meiosis I: Homologous chromosomes separate**

**Prophase I**
nuclear envelope fragment
crossing-over
spindle
sister chromatids
2n = 4
tetrad

**Metaphase I**
chromosome attached to a spindle fiber
centromere

**Anaphase I**
chromosomes still duplicated

**Telophase I**
cleavage furrow
n = 2

**Prophase I**
Tetrads form, and crossing-over occurs as chromosomes condense; the nuclear envelope fragments.

**Metaphase I**
Tetrads align at the spindle equator. Either homologue can face either pole.

**Anaphase I**
Homologues separate, and dyads move to poles.

**Telophase I**
Daughter nuclei are haploid, having received one duplicated chromosome from each homologous pair.

nucleus with either homologue of any other pair. In other words, each homologous pair separated independently of all other pairs.

Crossing-over as well as the independent assortment of chromosomes during meiosis I, shuffles the genes and increases genetic variation in resulting gametes.

During meiosis II (see Fig. 7.9), the centromeres divide and a haploid set of daughter chromosomes moves toward each pole of the spindle. At the completion of meiosis II, there are four daughter cells—each with the haploid number of chromosomes.

It is important for a gamete to have one member from each pair of homologous chromosomes because only in that way can a gamete pass on a copy of each kind of gene to the offspring.

### Figure 7.9 Meiosis II.

During meiosis II, daughter chromosomes consisting of one chromatid each move to the poles. Following meiosis II, there are four haploid daughter cells. Comparing the number of centromeres in the daughter cells with the number in the parent cell at the start of meiosis I verifies that the daughter cells are haploid. The daughter cells are genetically dissimilar from each other and from the parent cell.

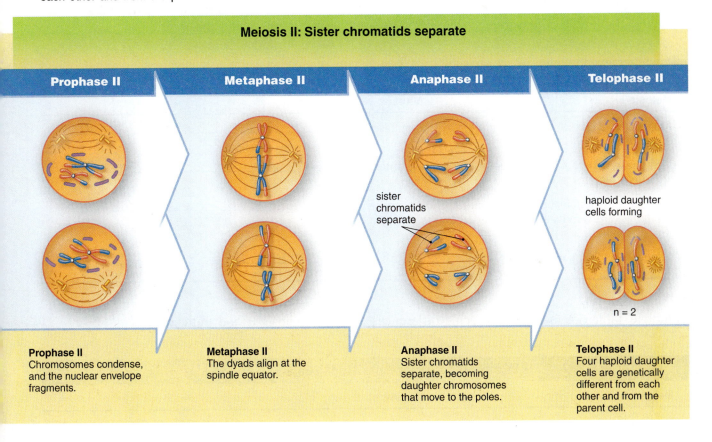

**Meiosis II: Sister chromatids separate**

**Prophase II**
Chromosomes condense, and the nuclear envelope fragments.

**Metaphase II**
The dyads align at the spindle equator.

**Anaphase II**
Sister chromatids separate, becoming daughter chromosomes that move to the poles.

**Telophase II**
Four haploid daughter cells are genetically different from each other and from the parent cell.

# 7.3 Mitosis Versus Meiosis

Examine Figure 7.10, and note the differences between mitosis and meiosis. Given that a parental cell is diploid (2n), fill in Table 7.2 to indicate general differences between mitosis and meiosis.

| Table 7.2 | Differences Between Mitosis and Meiosis | | |
|---|---|---|---|
| | | Mitosis | Meiosis |
| 1. | Number of divisions | | |
| 2. | Chromosome number in daughter cells | | |
| 3. | Number of daughter cells | | |

Complete Table 7.3 to indicate specific differences between meiosis I and mitosis.

| Table 7.3 | Meiosis I Compared with Mitosis |
|---|---|
| **Meiosis I** | **Mitosis** |
| Prophase I: Pairing of homologous chromosomes | Prophase: _____ |
| Metaphase I: Homologues paired at spindle equator | Metaphase: _____ |
| Anaphase I: Homologues of each pair separate, and duplicated chromosomes move to poles | Anaphase: _____ |
| Telophase I: Two haploid daughter cells | Telophase: _____ |

Complete Table 7.4 to indicate specific differences between meiosis II and mitosis.

| Table 7.4 | Meiosis II Compared with Mitosis |
|---|---|
| **Meiosis II** | **Mitosis** |
| Prophase I: No pairing of chromosomes | Prophase: _____ |
| Metaphase II: Haploid number of duplicated chromosomes at spindle equator | Metaphase: _____ |
| Anaphase II: Sister chromatids separate, becoming daughter chromosomes that move to the poles | Anaphase: _____ |
| Telophase II: Four haploid daughter cells, not genetically identical | Telophase: _____ |

## Figure 7.10  Meiosis compared to mitosis.

Compare metaphase I of meiosis to metaphase of mitosis. Only in metaphase I are the homologous chromosomes paired, forming tetrads, at the equator. Members of homologous chromosome pairs separate during anaphase I, and therefore the daughter cells are haploid. The blue chromosomes were inherited from the paternal parent, and the red chromosomes were inherited from the maternal parent. The exchange of color between nonsister chromatids represents the crossing-over that occurred during meiosis I.

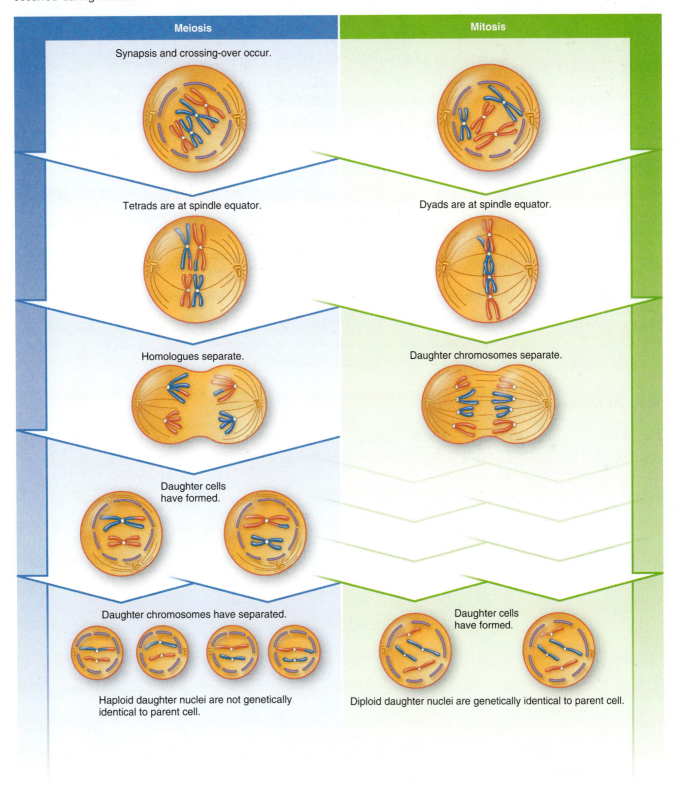

| Meiosis | Mitosis |
| --- | --- |
| Synapsis and crossing-over occur. | |
| Tetrads are at spindle equator. | Dyads are at spindle equator. |
| Homologues separate. | Daughter chromosomes separate. |
| Daughter cells have formed. | |
| Daughter chromosomes have separated. | Daughter cells have formed. |
| Haploid daughter nuclei are not genetically identical to parent cell. | Diploid daughter nuclei are genetically identical to parent cell. |

_____ 1. During what stage of the cell cycle does DNA replication occur?

_____ 2. Name the phase of cell division when separation of sister chromatids occurs.

_____ 3. By what process does the cytoplasm of a human cell separate?

_____ 4. Name the phase of cell division when duplicated chromosomes first appear.

_____ 5. Where in humans would you expect to find meiosis taking place?

_____ 6. If there are 13 pairs of homologous chromosomes in the parental cell, how many chromosomes are there in a sperm?

_____ 7. Cytokinesis in plant cells requires the formation of what structure?

_____ 8. A parental cell has two pairs of homologues. How many different combinations of chromosomes can occur among the gametes?

_____ 9. What do you call chromosomes that look alike and carry genes for the same traits?

_____ 10. If homologues are separating, what phase is this?

_____ 11. If the parent cell has 24 chromosomes, how many does each daughter cell have at the completion of meiosis II?

_____ 12. Name the type of cell division during which homologues pair.

_____ 13. Name the type of cell division described by 2n → 2n.

_____ 14. Does metaphase of mitosis, meiosis I, or meiosis II have the haploid number of chromosomes at the equator of the spindle?

## Thought Questions

15. A student is simulating meiosis I with homologous chromosomes that are red-long and yellow-long. Why would you not expect to find both red-long and yellow-long in one resulting daughter cell?

16. With reference to the same homologues, describe the appearance of two nonsister chromatids following crossing-over.

# 8

# Patterns of Inheritance

## Introduction

**Genes,** the units of heredity that control the specific characteristics of an individual, are arranged in a linear fashion along the chromosomes. Alternate forms of a gene having the same position (locus) on a pair of chromosomes and affecting the same trait are called **alleles** (Fig. 8.1). An individual can be homozygous dominant (two dominant alleles, *AA*), homozygous recessive (two recessive alleles, *aa*), or heterozygous (one dominant and one recessive allele, *Aa*). **Genotype** refers to an individual's genes, while **phenotype** refers to an individual's appearance. Homozygous dominant and heterozygous individuals show the dominant phenotype; homozygous recessive individuals show the recessive phenotype.

Of 23 pairs of chromosomes in each human cell, 22 are autosomes, and one pair is the sex chromosomes. The sex chromosomes carry genes just as the autosomes do. Some of these genes determine the sex of the individual (that is, whether the individual has testes or ovaries), but most of the genes on the sex chromosomes control traits unrelated to sexual characteristics. They are called sex-linked genes because they are on the sex chromosomes. Most of the sex-linked genes are on the X chromosomes.

**Figure 8.1** **Homologous chromosomes.**
The letters represent alleles— that is, alternate forms of a gene. Each allelic pair, such as *Gg* or *Tt,* is located on homologous chromosomes at a particular gene locus.

# 8.1 One-Trait Crosses

A single pair of alleles is involved in one-trait crosses. Reproduction between two heterozygous individuals (*Aa*) results in both dominant and recessive phenotypes among the offspring. The expected phenotypic ratio among the offspring from a **monohybrid cross** is 3:1. Three offspring have the dominant phenotype for every one that has the recessive phenotype.

These results are obtainable because the alleles of each parent separate during meiosis; otherwise, all offspring would inherit a dominant allele, and no offspring would be homozygous recessive.

Inheritance is a game of chance. Just as there is a 50% probability of heads or tails when tossing a coin, there is a 50% probability that a sperm or an egg will have an *A* or an *a* when the parent is *Aa*. The chance of an equal number of heads or tails improves as the number of tosses increases. In the same way, the chance of an equal number of gametes with *A* and *a* improves as the number of gametes increases. Therefore, the 3:1 ratio among offspring is more likely when a large number of sperm fertilize a large number of eggs.

## Color of Tobacco Seedlings

In tobacco plants, a dominant allele (*C*) for chlorophyll gives the plants a green color, and a recessive allele (*c*) for chlorophyll causes a plant to appear white. If a tobacco plant is homozygous for the recessive allele (*c*), it cannot manufacture chlorophyll and thus appears white (Fig. 8.2).

**Figure 8.2   Monohybrid cross.**
These tobacco seedlings are growing on an agar plate. The white plants cannot manufacture chlorophyll.

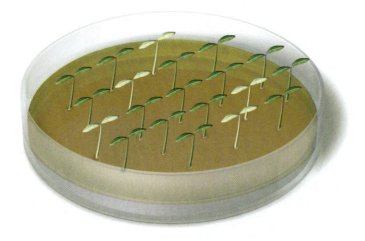

## Experimental Procedure: Color of Tobacco Seedlings

1. Obtain a numbered agar plate on which tobacco seedlings are growing. They are the offspring of the cross *Cc* × *Cc*. Complete this Punnett square:

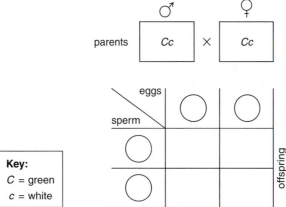

   What is the expected phenotypic ratio? _____

2. Using a binocular dissecting microscope, view the seedlings, and count the number that are green and the number that are white. Record the plate number and your results in Table 8.1.

3. Repeat steps 1 and 2 for two additional plates. Total the number that are green and the number that are white.

4. Complete Table 8.1 by recording the class data.

| Table 8.1 | Color of Tobacco Seedlings | |
|---|---|---|
| | *Number of Offspring* | |
| | **Green Color** | **White Color** |
| Plate # _____ | | |
| Plate # _____ | | |
| Plate # _____ | | |
| Totals | | |
| Class data | | |

### Conclusions

- Calculate the actual phenotypic ratio you observed. _____ Do your results differ from the expected ratio? _____ Explain. _____

  _____

- Repeat these steps using the class data. Do your class data give a ratio closer to the expected ratio? _____ Explain. _____

  _____

## 8.2 Human Inheritance

Human beings are subject to the same laws of genetics as are tobacco seedlings, corn, and fruit flies. Figure 8.3 illustrates some traits that we will be considering.

**Figure 8.3** **Commonly inherited characteristics in human beings.**
The alleles indicate which characteristics are dominant and which are recessive.

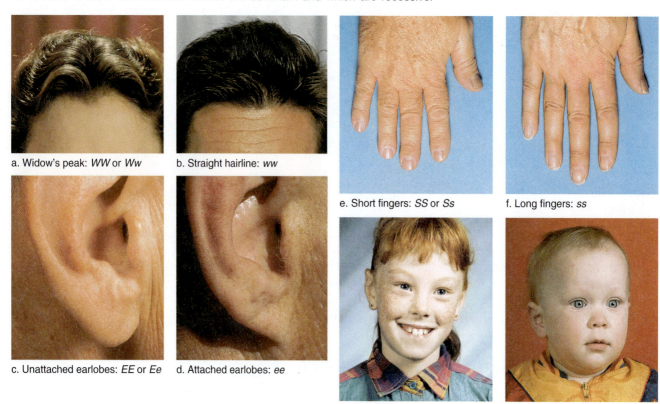

a. Widow's peak: *WW* or *Ww*

b. Straight hairline: *ww*

c. Unattached earlobes: *EE* or *Ee*

d. Attached earlobes: *ee*

e. Short fingers: *SS* or *Ss*

f. Long fingers: *ss*

g. Freckles: *FF* or *Ff*

h. No freckles: *ff*

### Autosomal Dominant and Recessive Traits

The alleles for autosomal traits are carried on the nonsex chromosomes. If individuals are homozygous dominant (*AA*) or heterozygous (*Aa*), their phenotype is the dominant trait. If individuals are homozygous recessive (*aa*), their phenotype is the recessive trait.

#### Experimental Procedure: Human Traits

1. For this Experimental Procedure, you will need a lab partner to help you determine your phenotype for the traits listed in the first column of Table 8.2.
2. Determine your probable genotype. If you have the recessive phenotype, you know your genotype. If you have the dominant phenotype, you may be able to decide whether you are homozygous dominant or heterozygous by recalling the phenotype of your parents, siblings, or children. Circle your probable genotype in the second column of Table 8.2.
3. Your instructor will tally the class's phenotypes for each trait so that you can complete the third column of Table 8.2.
4. Complete Table 8.2 by calculating the percentage of the class with each characteristic. Are dominant phenotypes always the most common in a population? _____ Explain. _____

Table 8.2 | Autosomal Human Traits

| Trait:  d = Dominant   r = Recessive | Possible Genotypes | Number in Class | Percentage of Class with Trait |
|---|---|---|---|
| **Hairline:** | | | |
| Widow's peak (d) | *WW* or *Ww* | | _____ |
| Straight hairline (r) | *ww* | | |
| **Earlobes:** | | | |
| Unattached (d) | *EE* or *Ee* | | _____ |
| Attached (r) | *ee* | | |
| **Skin pigmentation:** | | | |
| Freckles (d) | *FF* or *Ff* | | _____ |
| No freckles (r) | *ff* | | |
| **Hair on back of hand:** | | | |
| Present (d) | *HH* or *Hh* | | _____ |
| Absent (r) | *hh* | | |
| **Thumb hyperextension—"hitchhiker's thumb":** | | | |
| Last segment cannot be bent backward (d) | *TT* or *Tt* | | _____ |
| Last segment can be bent back to 60° (r) | *tt* | | |
| **Bent little finger:** | | | |
| Little finger bends toward ring finger (d) | *LL* or *Ll* | | _____ |
| Straight little finger (r) | *ll* | | |
| **Interlacing of fingers:** | | | |
| Left thumb over right (d) | *II* or *Ii* | | _____ |
| Right thumb over left (r) | *ii* | | |

### Genetics Problems

1. Nancy and the members of her immediate family have attached earlobes. Her maternal grandfather has unattached earlobes. What is the genotype of her maternal grandfather? _____ Nancy's maternal grandmother is no longer living. What could have been the genotype of her maternal grandmother? _____

2. Joe does not have a bent little finger, but his parents do. What is the expected phenotypic ratio among the parents' children? _____

3. Henry is adopted. He has hair on the back of his hand. Could both of his parents have had hair on the back of the hand? _____ Could both of his parents have had no hair on the back of the hand? _____ Explain. _____

# 8.3 Two-Trait Crosses

Two-trait crosses involve two pairs of alleles. During a **dihybrid cross,** when two dihybrid individuals (*AaBb*) reproduce, the expected phenotypic ratio among the offspring is 9:3:3:1, representing four possible phenotypes. These results are obtainable because the alleles of the parents separate independently of one another when the gametes are formed.

## Color and Texture of Corn

In corn plants, the allele for purple kernel (*P*) is dominant over the allele for yellow kernel (*p*), and the allele for smooth kernel (*S*) is dominant over the allele for rough kernel (*s*) (Fig. 8.4).

**Figure 8.4** **Dihybrid cross.**
Four types of kernels are seen on an ear of corn following a dihybrid cross: purple smooth, purple rough, yellow smooth, and yellow rough.

20 μm

---

### Experimental Procedure: Color and Texture of Corn

1. Obtain an ear of corn from the supply table. You will be examining the results of the cross *PpSs* × *PpSs*. Complete this Punnett square:

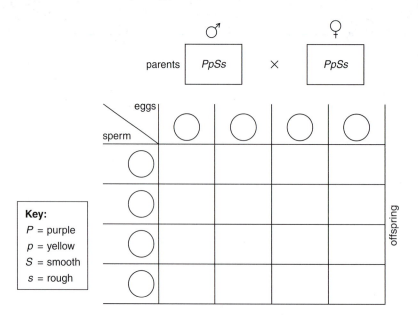

**Key:**
*P* = purple
*p* = yellow
*S* = smooth
*s* = rough

What is the expected phenotypic ratio among the offspring? _____

2. Count the number of kernels of each possible phenotype listed in Table 8.3. Record the sample number and your results in Table 8.3. Use three samples, and total your results for all samples. Also record the class data.

| Table 8.3 | Color and Texture of Corn | | | |
|---|---|---|---|---|
| | Number of Kernels | | | |
| | Purple Smooth | Purple Rough | Yellow Smooth | Yellow Rough |
| Sample # _____ | | | | |
| Sample # _____ | | | | |
| Sample # _____ | | | | |
| Totals | | | | |
| Class data | | | | |

## Conclusions

- From the data, which two traits seem dominant? _____ and _____
  Which two traits seem recessive? _____ and _____
- Calculate the actual phenotypic ratio you observed. _____ Do your results differ from the expected ratio? _____
- Repeat these steps using the class data. Do your class data give a ratio closer to the expected ratio? _____ Explain. _____

  _____

- What cross would you do to test if a plant that produces only purple, smooth kernels is heterozygous or homozygous? _____ What would be the resulting phenotypic ratio if the plant is heterozygous? _____ Homozygous? _____
  Use the following space to determine the results by using Punnett squares.

# 8.4 X-Linked Crosses

The sex chromosomes carry genes that affect traits other than the individual's sex. Genes on the sex chromosomes are called **sex-linked genes.** The vast majority of sex-linked genes have alleles on the X chromosome and are called **X-linked genes.** Most often, an abnormal X-linked condition is recessive.

Color blindness is an X-linked, recessive trait. The possible genotypes and phenotypes are as follows:

**Females**

$X^B X^B$ = normal vision

$X^B X^b$ = normal vision

$X^b X^b$ = color blind

**Males**

$X^B Y$ = normal vision

$X^b Y$ = color blind

## Experimental Procedure: X-Linked Traits

1. Your instructor will provide you with a color blindness chart. Have your lab partner present the chart to you. Write down the words or symbols you see, but do not allow your partner to see what you write, and do not discuss what you see. This is important because color-blind people see something different than do people who are not color blind.
2. Now test your lab partner as he or she has tested you.
3. Are you color blind? _____ If so, what is your genotype? _____
4. If you are a female and are not color blind, you can judge whether you are homozygous or heterozygous by knowing if any member of your family is color blind. If your father is color blind, what is your genotype? _____ If your mother is color blind, what is your genotype? _____ If you know of no one in your family who is color blind, what is your probable genotype? _____

### Genetics Problems

1. The only color-blind member of Arlene's family is her brother. What is her brother's genotype? _____ What is her father's genotype? _____ What is her mother's genotype? _____ What is Arlene's genotype if she later has a color-blind son? _____
2. Bill's grandfather and brother are color blind. Bill is not color blind. Why not?

_____

## Drosophila melanogaster

It is possible to do an X-linked cross by using the fruit fly *Drosophila melanogaster*. First, your instructor may provide you with a culture bottle that contains these small flies. Both the adults and the larvae of *Drosophila melanogaster* feed on plant sugars and on the wild yeasts that grow on rotting fruit. The female flies first lay **eggs** on the same materials. After a day or two, the eggs hatch into small **larvae** that feed and grow for about eight days, depending on the temperature. During this period, they **molt** twice. Therefore, three larval periods of growth, called **instars,** occur between molts. When fully grown, the third-stage instar larvae cease feeding and **pupate.** During pupation, which lasts about four days, the larval tissues are reorganized to form those of the adult. The life cycle is summarized in Figure 8.5.

## Figure 8.5 Life cycle of *Drosophila.*

The adult female lays eggs that go through three larval stages before pupating. Metamorphosis, which occurs during pupation, produces the adult fly.

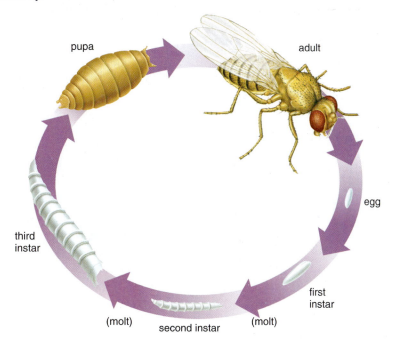

## Observation: Drosophila melanogaster

You will be provided with slides, frozen flies, or live flies to examine. If the flies are alive, follow the directions provided by your instructor for using FlyNap™ to anesthetize them.

1. Put frozen or anesthetized flies on a white card, and use a camel-hair brush to move them around. Use a binocular dissecting microscope or a hand lens to see the flies clearly.
2. If you are looking at slides, you may use the scanning lens of your compound light microscope to view the flies.
3. Examine wild-type flies. Complete Table 8.4 for wild-type flies. Long wings extend beyond the body, and vestigial (short) wings do not extend beyond the body.
4. Examine mutant flies. Complete Table 8.4 for all the mutant flies you examined.

| Table 8.4 | Characteristics of Wild-Type and Mutant Flies | | | | |
|---|---|---|---|---|---|
| | **Wild-Type** | **Ebony Body** | **Vestigial-Wing** | **Sepia-Eye** | **White-Eye** |
| Wing length | | | | | |
| Color of eyes | | | | | |
| Color of body | | | | | |

5. Use the following characteristics to distinguish male flies from female flies (Fig. 8.6):

- The male is generally smaller.
- The male has a more rounded abdomen than the female. The female has a pointed abdomen.
- The male has sex combs on the forelegs.
- Dorsally, the male is seen to have a black-tipped abdomen, whereas the female appears to have dark lines only at the tip.
- Ventrally, the abdomen of the male has a dark region at the tip due to the presence of claspers; this dark region is lacking in the female.

**Figure 8.6** *Drosophila* **male versus** *Drosophila* **female.**

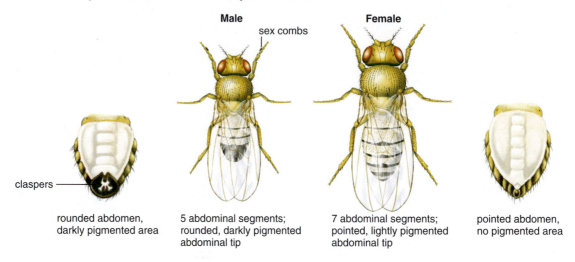

Male — sex combs

Female

claspers

rounded abdomen, darkly pigmented area

5 abdominal segments; rounded, darkly pigmented abdominal tip

7 abdominal segments; pointed, lightly pigmented abdominal tip

pointed abdomen, no pigmented area

## Red/White Eye Color in *Drosophila*

In fruit flies, red eyes ($X^R$) are dominant over white eyes ($X^r$). You will be examining the results of the cross $X^R X^r \times X^R Y$. Complete this Punnett square:

**Key:**
$X^R$ = red eyes
$X^r$ = white eyes

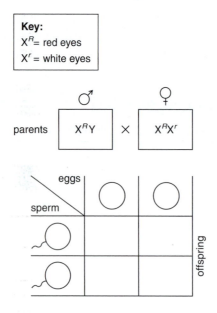

parents ♂ $X^R Y$ × ♀ $X^R X^r$

eggs

sperm

offspring

What are the expected phenotypic results of this cross?

Females: _____ Males: _____

## Experimental Procedure: Red/White Eye Color in *Drosophila*

The cross described here will take three weeks. The experiment can be done in one week if your instructor provides you with a vial that already contains the results of the cross $X^R X^r \times X^R Y$. In this case, proceed directly to step 3.

1. **Week 1:** Place the parent flies ($X^R X^r \times X^R Y$) in a prepared culture vial. Your instructor will show you how to use instant medium and dry yeast to prepare the vial. Label your culture.

   What is the phenotype of the female and male flies you are using? _____

   What is the genotype of the female flies? _____

   What is the genotype of the male flies? _____

2. **Week 2:** Remove the parent flies from the vial before their offspring pupate.

   Why is it necessary to remove these flies before you observe your results? _____

   _____

3. **Week 3:** Observe the results of the cross by counting the offspring. Follow directions provided by your instructor for anesthetizing and removing flies. When counting, use the binocular dissecting microscope or a hand lens. Divide your flies into the following groups: (1) red-eyed males, (2) red-eyed females, (3) white-eyed males, and (4) white-eyed females. Record your results in Table 8.5. Also record the class data.

| Table 8.5 | Red/White Eye Color in *Drosophila* | |
|---|---|---|
| | *Number of Offspring* | |
| **Your Data** | **Red Eyes** | **White Eyes** |
| Males | | |
| Females | | |
| Class data<br>  Males | | |
|   Females | | |

## Conclusions

- Calculate the actual phenotypic ratio you observed for males and females separately.

  Males: _____

  Females: _____

- Do your results differ from the expected ratio? _____
- Repeat these steps using the class data. Do your class data give a ratio closer to the

  expected ratio? _____ Explain. _____

  _____

_____ 1. A cross gives a 3:1 phenotypic ratio. What are the genotypes of the parents?

_____ 2. Parents who both have the genotype *Aa* would produce what type of gametes?

_____ 3. What is the genotype of a man who has unattached earlobes but whose mother has attached earlobes?

_____ 4. A wild-type *Drosophila* fly has long wings, gray body, and what color eyes?

_____ 5. When doing a genetic cross, why is it necessary to remove the parent flies before the pupae have hatched?

_____ 6. How many different types of gametes would an *AaBb* parent have?

_____ 7. What is the phenotypic ratio among offspring if both parents are dihybrids?

_____ 8. Why do you expect class data to be closer to the expected ratio than your individual data?

_____ 9. What is the genotype of a white-eyed male fruit fly?

_____ 10. Which gender can have white eyes if the female parent is homozygous dominant for red eyes and if the male parent has white eyes?

_____ 11. Why don't you expect to get a ratio that is exactly 3:1 for a monohybrid cross?

## Thought Questions

12. What are the results of the cross *AaBb* × *aaBb*?

13. You count 73 long-winged flies and 27 vestigial-winged flies from a cross between two heterozygous parents. How many flies would you have expected to have long wings?

# 9

# DNA Biology and Technology

## Learning Objectives

**9.1 DNA Structure and Replication**
- Explain how the structure of DNA facilitates replication.
- Explain why DNA replication is semiconservative.

**9.2 RNA Structure**
- List the ways in which RNA structure differs from DNA structure.

**9.3 DNA and Protein Synthesis**
- State the function of transcription and translation during protein synthesis.
- Explain how DNA stores information.

**9.4 Isolation of DNA**
- Describe how DNA can be isolated, and explain the procedure for testing DNA.

**9.5 Polymerase Chain Reaction (PCR)**
- Explain the purpose of the PCR.
- List some applications of PCR and explain the process of DNA fingerprinting.

## Introduction

This laboratory pertains to molecular genetics and biotechnology. Molecular genetics is the study of the structure and function of **DNA (deoxyribonucleic acid),** the genetic material. Biotechnology is the manipulation of DNA for the benefit of human beings and other organisms.

First we will study the structure of DNA and see how that structure facilitates DNA replication in the nucleus of cells. DNA replicates prior to cell division; following cell division, each daughter cell has a complete copy of the genetic material. DNA replication is also needed to pass genetic material from one generation to the next. You may have an opportunity to use models to see how replication occurs.

Then we will study the structure of **RNA (ribonucleic acid)** and how it differs from that of DNA, before examining how DNA with the help of RNA specify protein synthesis. The linear construction of DNA, in which nucleotide follows nucleotide, is paralleled by the linear construction of the primary structure of protein, in which amino acid follows amino acid. Essentially, we will see that the sequence of nucleotides in DNA codes for the sequence of amino acids in a protein. We will also review the role of three types of RNA in protein synthesis. DNA's code is passed to messenger RNA (mRNA), which moves to the ribosomes containing ribosomal RNA (rRNA). Transfer RNA (tRNA) brings the amino acids to the ribosomes and they become sequenced in the order directed by mRNA.

You will have an opportunity in this laboratory to isolate DNA from a vegetable or fruit. All organisms use DNA to code for the sequence of amino acids in proteins. This constitutes biochemical evidence that all living things are related to the same original ancestor.

# 9.1 DNA Structure and Replication

The structure of DNA lends itself to **replication,** the process that makes a copy of a DNA molecule. DNA replication is a necessary part of chromosome duplication, which precedes cell division. It also makes possible the passage of DNA from one generation to the next.

## DNA Structure

DNA is a polymer of nucleotide monomers (Fig. 9.1). Each nucleotide is composed of three molecules: deoxyribose (a 5-carbon sugar), a phosphate, and a nitrogen-containing base.

**Figure 9.1** **Overview of DNA structure.**

*a.* Complementary base pairing dictates that A is bonded to T and G is bonded to C and vice versa. The two strands of the molecule are antiparallel; that is, the sugar-phosphate groups are oriented in different directions: the 5′ end of one strand is opposite the 3′ end of the other strand. *b.* Diagram of a DNA double helix shows that the molecule resembles a twisted ladder. Sugar-phosphate backbones make up the sides of the ladder, and hydrogen-bonded bases make up the rungs of the ladder. *Label the figure as directed in the next Observation.*

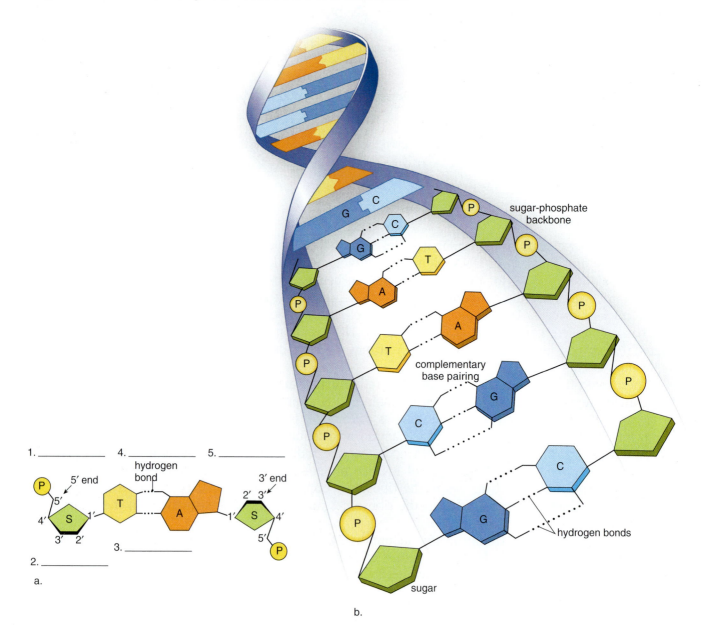

1.  A nucleotide pair is shown in Figure 9.1*a*. If you are working with a kit, draw a representation of one of your nucleotides here. *Label phosphate, base, and deoxyribose in your drawing and in Figure 9.1a, 1–5.*

2.  Notice the four types of bases: cytosine (C), thymine (T), adenine (A), and guanine (G). What is the color of the four types of bases in Figure 9.1*b*? In your kit? Complete Table 9.1 by writing in the colors of the bases.

| Table 9.1 | Base Colors | |
|-----------|-------------|-------------|
| | **In Figure 9.1b** | **In Your Kit** |
| Cytosine | | |
| Thymine | | |
| Adenine | | |
| Guanine | | |

3.  Using Figure 9.1*b* as a guide, join several nucleotides together. Observe the entire DNA molecule. What type of molecules make up the backbone (uprights of ladder) of DNA (Fig. 9.1*b*)?

    _____ and _____ In the backbone, the phosphate of one nucleotide is bonded to a sugar of the next nucleotide.

4.  Using Figure 9.1*b* as a guide, join the bases together with hydrogen bonds. Label a hydrogen bond in Figure 9.1*b*. Dots are used to represent hydrogen bonds in Figure 9.1*b* because hydrogen bonds are (strong or weak)? _____

5.  In Figure 9.1*b* and in your model, the base A is always paired with the base _____, and

    the base C is always paired with the base _____. This is called complementary base pairing.

6.  In Figure 9.1*b*, what molecules make up the rungs of the ladder? _____

7.  Each half of the DNA molecule is a DNA strand. Why is DNA also called a double helix

    (Fig. 9.1*b*)? _____

## DNA Replication

During replication, the DNA molecule is duplicated so that there are two DNA molecules. We will see that complementary base pairing makes replication possible.

1. Before replication begins, DNA is unzipped. Using Figure 9.2*a* as a guide, break apart your two DNA strands. What bonds are broken to unzip the DNA strands? _____ _____

2. Using Figure 9.2*b* as a guide, attach new complementary nucleotides to each strand using complementary base pairing.

3. Show that you understand complementary base pairing by completing Table 9.2. You now have two DNA molecules (Fig. 9.2*c*). Are your molecules identical? _____

4. Because of complementary base pairing, each new double helix is composed of an _____ strand and a _____ strand. *Write old or new beside each strand in Figure 9.2a, b, and c, 1–10. Conservative means to save something from the past.* Why is DNA replication called semiconservative?

**Figure 9.2** **DNA replication.**
Use of the ladder configuration better illustrates how replication takes place. *a.* The parental DNA molecule. *b.* The "old" strands of the parental DNA molecule have separated. New complementary nucleotides available in the cell are pairing with those of each old strand. *c.* Replication is complete.

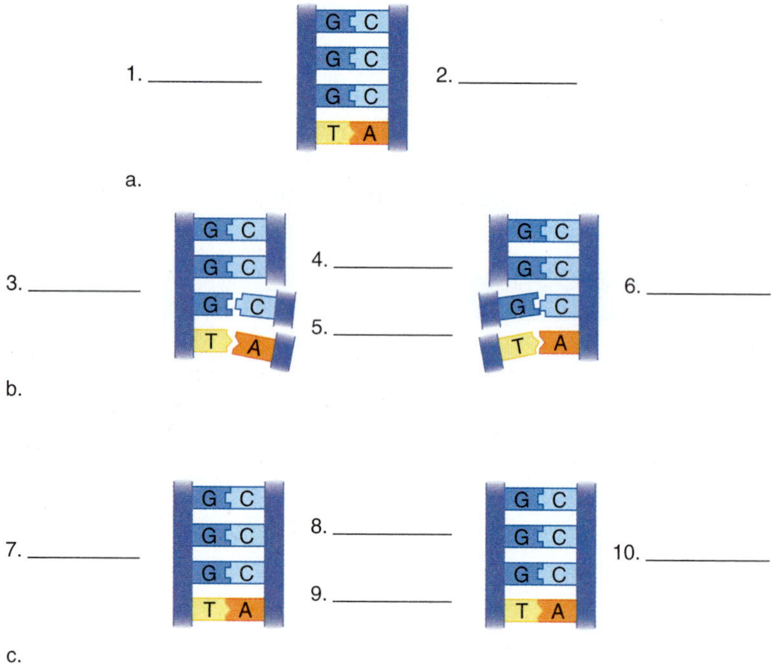

1. _____     2. _____

a.

3. _____     4. _____     6. _____
                 5. _____

b.

7. _____     8. _____     10. _____
                 9. _____

c.

_____
_____
_____
_____

5. Genetic material has to be inherited from cell to cell and organism to organism. Consider that because of DNA replication a chromosome is composed of two chromatids and each chromatid is a complete DNA molecule. The chromatids separate during cell division so that each daughter cell receives a copy of each chromosome. Does replication provide a means for passing DNA from cell to cell and organism to organism? _____
   Explain. _____
   _____
   _____

| Table 9.2 | DNA Replication |
|---|---|
| Old strand | G G G T T C C A T T A A A T T C C A G A A A T C A T A |
| New strand |  |

## 9.2 RNA Structure

Like DNA, RNA is a polymer of nucleotides (Fig. 9.3). In an RNA nucleotide, the sugar ribose is attached to a phosphate molecule and to a nitrogen-containing base, C, U, A, or G. In RNA, the base uracil replaces thymine as one of the pyrimidine bases. RNA is single stranded, whereas DNA is double stranded.

**Figure 9.3** Overview of RNA structure.

RNA is a single strand of nucleotides. *Label one nucleotide as directed in the next Observation.*

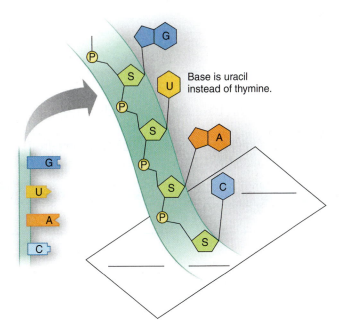

Base is uracil instead of thymine.

1. Describe the backbone of an RNA molecule. _____

2. Where are the bases located in an RNA molecule? _____

3. Complete Table 9.3 to show the complementary DNA bases for the RNA bases.

| Table 9.3 | DNA and RNA Bases | | | |
|---|---|---|---|---|
| RNA Bases | C | U | A | G |
| DNA Bases | | | | |

### Observation: RNA Structure

1. If you are using a kit, draw a nucleotide for the construction of mRNA. *Label the ribose (the sugar in RNA), the phosphate, and the base in your drawing and in Figure 9.3.*

2. Complete Table 9.4 by writing in the colors of the bases in Figure 9.3 and in your kit.

| Table 9.4 | Base Colors | |
|---|---|---|
| | In Figure 9.3 | In Your Kit |
| Cytosine | | |
| Uracil | | |
| Adenine | | |
| Guanine | | |

3. Why in Table 9.5 should you list the base uracil instead of the base thymine in RNA? _____

Complete Table 9.5 to show the several other ways RNA differs from DNA.

| Table 9.5 | DNA Structure Compared with RNA Structure | |
|---|---|---|
| | DNA | RNA |
| Sugar | Deoxyribose | |
| Bases | Adenine, guanine, thymine, cytosine | |
| Strands | Double stranded with base pairing | |
| Helix | Yes | |

4. Double-stranded DNA facilitates _____. We will see that single-stranded RNA facilitates protein synthesis.

## 9.3 DNA and Protein Synthesis

Protein synthesis requires the processes of transcription and translation. During **transcription,** which takes place in the nucleus, an RNA molecule called **messenger RNA (mRNA)** is made complementary to one of the DNA strands. This mRNA leaves the nucleus and goes to the ribosomes in the cytoplasm. Ribosomes are composed of **ribosomal RNA (rRNA)** and proteins in two subunits.

During **translation,** RNA molecules called **transfer RNA (tRNA)** bring amino acids to the ribosome, and they join in the order prescribed by mRNA.

In the end, the final sequence of amino acids in a protein is specified by DNA. This is the information that DNA, the genetic material, stores.

## Transcription

During transcription, complementary RNA is made from a DNA template (Fig. 9.4). A portion of DNA unwinds and unzips at the point of attachment of the enzyme RNA polymerase. A strand of mRNA is produced when complementary nucleotides join in the order dictated by the sequence of bases in DNA. Transcription occurs in the nucleus, and the mRNA passes out of the nucleus to enter the cytoplasm.

*Label Figure 9.4. For number 1, name this DNA strand. For number 2, note the name of the enzyme that carries out mRNA synthesis. For number 3, note where this molecule will be active.*

### Observation: Transcription

1. If you are using a kit, unzip your DNA model so that only one strand remains. This strand is the **template strand,** the strand that is transcribed.
2. Using Figure 9.4 as a guide, construct a messenger RNA (mRNA) molecule by first lining up RNA nucleotides that are complementary to the sense strand of your DNA molecule. Join the nucleotides together to form mRNA.
3. A portion of DNA has the sequence of bases shown in Table 9.6. *Complete Table 9.6 to show the sequence of bases in mRNA.*
4. If you are using a kit, unzip mRNA transcript from the DNA. Locate the end of the strand that will move into the cytoplasm.

**Figure 9.4** **Messenger RNA (mRNA).**
Messenger RNA complementary to a section of DNA forms during transcription.

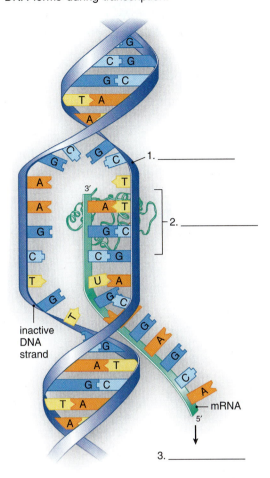

1. _____

2. _____

inactive DNA strand

mRNA

3. _____

| Table 9.6 | Transcription | | | | | | | | | | | | | | | | | |
|---|---|---|---|---|---|---|---|---|---|---|---|---|---|---|---|---|---|---|
| DNA | T | A | C | A | C | G | A | G | C | A | A | C | T | A | A | C | A | T |
| mRNA | | | | | | | | | | | | | | | | | | |

### Conclusion

- What is the function of transcription? _____

_____

## Translation

DNA specifies the sequence of amino acids in a polypeptide because every three bases stands for an amino acid. Therefore, DNA is said to have a **triplet code.** The bases in mRNA are complementary to those in DNA, and therefore, every three bases in mRNA (called a **codon**) stands for the same sequence of amino acids as does DNA. The correct sequence of amino acids in a polypeptide is the message that mRNA carries.

Messenger RNA leaves the nucleus and proceeds to the ribosomes, where protein synthesis occurs. Transfer RNA (tRNA) molecules are so named because they transfer amino acids to the ribosomes. Each RNA has a specific tRNA amino acid at one end and a matching **anticodon** at the other end (Fig. 9.5). *Label Figure 9.5 (1–3),* where the amino acid is represented as a colored ball, the tRNA is yellow, and the anticodon is the sequence of three bases. (The anticodon is complementary to the mRNA codon.)

**Figure 9.5** **Transfer RNA (tRNA).**
Transfer RNA carries amino acids to the ribosomes.

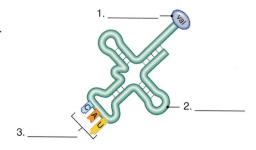

1. _____
2. _____
3. _____

*Observation: Translation*

1. Figure 9.6 shows seven tRNA-amino acid complexes. Every amino acid has a name; in the figure, only the first three letters of the name are inside the ball. Using the mRNA sequence given in Table 9.7, number the tRNA–amino acid complexes in the order they will come to the ribosome.

2. If you are using a kit, arrange your tRNA–amino acid complexes in the proper order. Complete Table 9.7. Why are the codons and anticodons in groups of three? _____

**Figure 9.6** **Transfer RNA diversity.**
Each type of tRNA carries only one particular amino acid, designated here by the first three letters of its name.

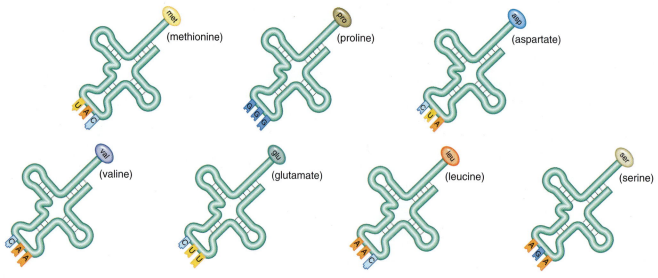

(methionine)   (proline)   (aspartate)

(valine)   (glutamate)   (leucine)   (serine)

| Table 9.7 | Translation | | | | | | |
|-----------|-------------|-----|-----|-----|-----|-----|-----|
| mRNA codons | | AUG | CCC | GAU | GUU | GAA | UUG | UCU |
| tRNA anticodons | | | | | | | | |
| Amino acid* | | | | | | | | |

*Use three letters only. See Figure 9.6 for the full names of these amino acids.

3. Figure 9.7 shows the manner in which the polypeptide grows. A ribosome has room for two tRNA complexes at a time; one is at the P site and the other is at the A site. The tRNA at the P site passes its peptide to the tRNA-amino acid complex at the A site. Then, the tRNA at the P site leaves. The ribosome moves forward (translocation), and the tRNA-peptide complex is now at the P site. A new tRNA-amino acid complex comes to the A site. This sequence of events occurs over and over until the entire polypeptide is borne by the last tRNA to come to the ribosome. *Write a description for events 2–4 in Figure 9.7.*

**Figure 9.7** Protein synthesis.
1. A ribosome has room for two tRNA complexes at a time; one is at the P site and the other is at the A site. Three events occur over and over again during polypeptide synthesis. Describe these events at 2–4.

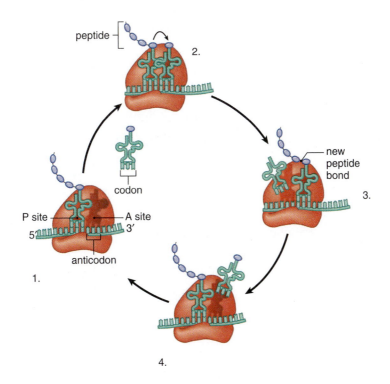

## 9.4 Isolation of DNA

In the following Experimental Procedure, you will isolate DNA from the cells of an organism using a modified procedure like that used worldwide in biotechnology laboratories. You will extract DNA from a vegetable or fruit filtrate that contains DNA in solution. To prepare the filtrate, your instructor homogenized the vegetable or fruit with a detergent. The detergent emulsifies and forms complexes with the lipids and proteins of the plasma membrane, causing them to precipitate out of solution. Cell contents, including DNA, become suspended in solution. The cellular mixture is then filtered to produce the filtrate that contains DNA and its adhering proteins.

The DNA molecule is easily degraded (broken down), so it is important to follow all instructions closely. Handle glassware carefully to prevent nucleases in your skin from contaminating the glassware.

### Experimental Procedure: Isolating DNA

1. Obtain a pair of gloves and wear them when doing this procedure.
2. Obtain a large, clean test tube and place it in an ice bath. Let stand for a few minutes to make sure the test tube is cold. Everything must be kept very cold.
3. Obtain approximately 4 ml of the *filtrate*, and add it to your test tube while keeping the tube in the ice bath.
4. Obtain and add 2 ml of cold *meat tenderizer solution* to the solution in the test tube, and mix the contents slightly with a stirring rod or Pasteur pipette. Let stand for 10 minutes so the enzyme has time to strip the DNA of protein.
5. Use a graduated cylinder or pipette to slowly add an equal volume (approximately 6 ml) of ice-cold *95% ethanol* along the inside of the test tube. Keep the tube in the ice bath, and tilt it to a 45° angle. You should see a distinct layer of ethanol over the white precipitate, the DNA. Let the tube sit for 2 to 3 minutes.
6. Insert a glass rod or a Pasteur pipette into the tube until it reaches the bottom of the tube. *Gently* swirl the glass rod or pipette, always in the same direction. (You are not trying to mix the two layers; you are trying to wind the DNA onto the glass rod like cotton candy.) This process is called "spooling" the DNA. The stringy, slightly gelatinous material that attaches to the pipette is DNA (Fig. 9.8). If the DNA has been damaged, it will still precipitate, but as white flakes that cannot be collected on the glass rod.
7. Answer the following questions:

   a. This procedure requires homogenization. When did homogenization occur? _____
      What was the purpose of homogenization? _____
   b. Next, deproteinization stripped proteins from the DNA. Which of the preceding steps represents deproteinization? _____
   c. Finally, DNA was precipitated out of solution. Which of the preceding steps represents the precipitation of DNA? _____

**Figure 9.8** **Isolation of DNA.**
The addition of ethanol causes DNA to come out of solution so that it can be spooled onto a glass rod.

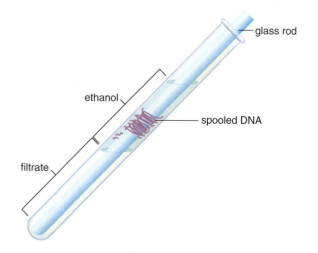

## 9.5 Polymerase Chain Reaction (PCR)

The **polymerase chain reaction (PCR)** creates millions of copies of a segment of DNA very quickly in a test tube. PCR can seek out and duplicate a segment of DNA from human blood, hair, or tissue specimens (even if they are millions of years old) and from microbes, other animals, or plants. It takes its name from the enzyme that copies the DNA segment over and over. The name of the enzyme is *DNA polymerase*. To copy a portion of the DNA, the enzyme needs a supply of DNA nucleotides, the monomers for a DNA molecule. PCR also requires primers that flank the targeted DNA—that is, the segment of DNA to be copied. *Complete Figure 9.9 by writing in the substances that need to be added to a test tube to carry out PCR.*

**Figure 9.9  Polymerase chain reaction.**
At the end of the first cycle, there are two identical copies of double-stranded DNA. Usually 25–30 cycles are run in an automated cycler, resulting in over 30 million copies of the original DNA double strand.

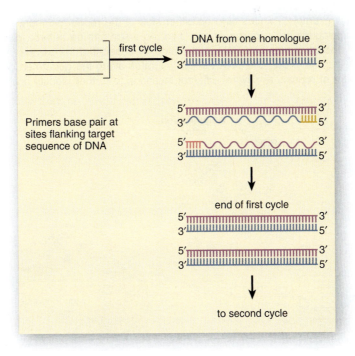

PCR has many applications. PCR can assist (1) in the diagnosis of a disease because it can detect the presence of pathogen DNA (virus or bacteria) in a sample of body fluids; (2) in detecting a mutation signifying a person has or is a carrier for a particular genetic disorder or a mutation that signifies a person has a particular type of cancer; and (3) in carrying out DNA fingerprinting for the purpose of identifying a particular individual. DNA fingerprinting can identify a person's relative, a deceased person, or a person who has committed a crime.

### DNA Fingerprinting

A **DNA fingerprint** is a unique difference in an individual's DNA that can be detected by the use of a gel electrophoresis. **Gel electrophoresis** is the migration of a DNA (or protein) sample through a gel in the presence of an electric field. The genome (all the genetic material) of an individual contains portions of DNA (i.e., genes) that code for the various proteins. Other DNA portions are sometimes called "junk" because they do not code for proteins—these sections are simply repeats of the same short sequence of bases over and over as in ATCATCATCATC.

Today's method of constructing a DNA fingerprint makes use of the observation that people differ by length of DNA repeats. (People differ because the number of repeats is inherited.) The greater the number of repeats, the greater the segment of DNA copied by PCR.

During gel electrophoresis, a sample of DNA is placed at one end (in a well) of a slab of gelatin and electricity is used to provide a negative and positive pole. A DNA molecule migrates from the negative pole to the positive pole because DNA is negatively charged. Small segments are faster moving than larger segments; therefore, DNA segments are separated according to their size. Figure 9.10 shows the results of PCR followed by gel electrophoresis in three different individuals when the same segment of DNA was copied. In number 3 in Figure 9.10, what do the small numbers refer to?

Why would an individual have, say, 20 repeats on one homologue and 18 repeats on the other homologue?

**Figure 9.10** **Use of PCR to perform DNA fingerprinting.**
People differ by the number of base repeats in particular regions of DNA. Therefore, it is possible to use these regions to do DNA fingerprinting as in this example. Primers bind on either side of the repeat segment before amplification begins. Strands of DNA from both homologues are necessarily amplified; a person can be heterozygous for the number of repeats as in individual I and III. The amount of amplified DNA is detected by separating the sample of DNA by gel electrophoresis. The lower the number of repeats, the farther the DNA sample migrates along the gel. When doing a DNA fingerprint, it is customary to detect the number of repeats at several different locations to further identify one individual from another.

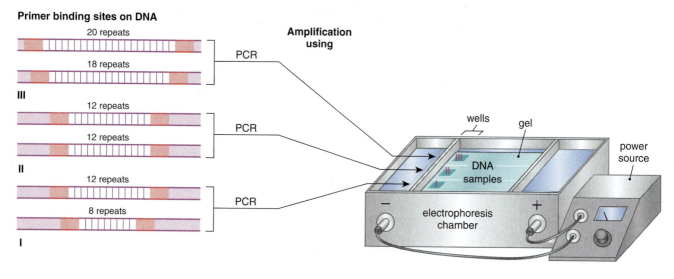

1. Use PCR to detect number of repeats in DNA.

2. Apply DNA samples to gel and perform electrophoresis.

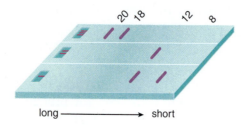

3. Analyze the DNA patterns.

1. A laboratory technician is provided with a sample of crime scene DNA (CS) and DNA samples for three suspects ($S_1$, $S_2$, and $S_3$). The laboratory technician prepares four DNA fingerprints.
2. *In Figure 9.11, using the lines provided, label the location of the wells, the negative pole, and the positive pole. Place an arrow on the left-hand side showing the direction of migration by DNA samples.*

## Figure 9.11 DNA fingerprints.

These are the DNA fingerprints for the crime scene DNA sample and the three suspects' samples. In this case, comparing DNA fragment patterns allows you to determine who committed the crime. In other instances, DNA fragment patterns allow you to determine who is the parent of a child, whether a person has a genetic disorder, or the identify of the remains following death.

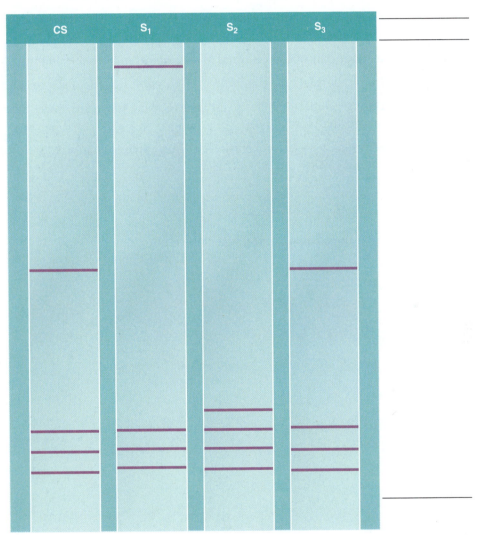

*Conclusion*

- Which suspect committed the crime? _____ How do you know? _____

_____

_____  1. The DNA structure resembles a twisted ladder. What molecules make up the sides of the ladder?

_____  2. What makes up the rungs of the ladder?

_____  3. Do the two DNA double helices following DNA replication have the same, or a different, composition?

_____  4. If DNA has 20% of adenine bases, what would be the percentage of thymine?

_____  5. If the codons are AUG, CGC, and UAC, what are the anticodons?

_____  6. Where does protein synthesis take place?

_____  7. What type of base pairing occurs during transcription?

_____  8. During transcription, what type of RNA is formed to take DNA's message to the ribosomes?

_____  9. In what part of the cell does translation occur?

_____  10. During translation, what type of RNA carries amino acids to the ribosomes?

_____  11. During gel electrophoresis, do long or short fragments travel more quickly toward the positive pole?

_____  12. Following gel electrophoresis of DNA samples, what was the resulting DNA pattern called?

## Thought Questions

13. What role does mRNA play in transcription and translation?

14. Explain the manner in which DNA fingerprinting identifies an individual.

# 10

# Genetic Counseling

## Learning Objectives

**10.1 Chromosome Inheritance**
- Understand the preparation and significance of a karyotype.
- Explain how nondisjunction occurs and what numerical sex chromosome abnormalities may result from nondisjunction.

**10.2 Genetic Disorders: The Present**
- Determine whether a pedigree represents a pattern of autosomal dominant, autosomal recessive, or X-linked recessive inheritance.
- Solve genetics problems involving autosomal dominant, autosomal recessive, and X-linked recessive alleles.
- Know what type of testing is available for genetic disorders and what options are available to prospective parents that have a genetic disorder.

**10.3 Genetic Disorders: The Future**
- Understand the relationships between abnormal DNA base sequences and genetic disorders.

## Introduction

Chromosomal inheritance has a marked effect on the general anatomy and physiology of the individual. If by chance the individual has an abnormality in chromosome number or structure, a syndrome results. A **syndrome** is a group of symptoms that appear together and tend to indicate the presence of a particular disorder.

**Genes,** the units of heredity located on the chromosomes, determine whether an individual has a genetic disorder. Individuals do not have an autosomal recessive disorder unless they have inherited two mutated alleles. For an autosomal dominant disorder, one mutated allele is sufficient. Most sex-linked disorders are carried on the X chromosome, and males receive only one X chromosome; therefore, they will display an X-linked recessive disorder even if they receive only a single mutated allele.

We now understand that a mutated gene has an altered DNA base sequence. For example, in cystic fibrosis (Fig. 10.1), DNA codes for a sequence of amino acids in a particular membrane protein; this protein is abnormal and usually has a missing amino acid. This results in an altered shape, which entraps the chloride ions inside cells. Water entering these cells leaves behind a very thick mucus, which clogs the bronchi and bronchioles, resulting in respiratory infections. Other organs in the body are also affected.

**Figure 10.1** **Cause of cystic fibrosis.**

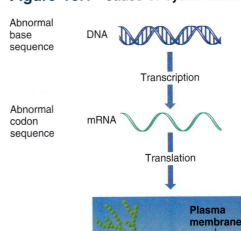

Abnormal base sequence — DNA

Transcription

Abnormal codon sequence — mRNA

Translation

Abnormal amino acid sequence results in a protein that is unable to function properly.

Plasma membrane

cystic fibrosis transmembrane conductance regulator (CFTR) protein

# 10.1 Chromosome Inheritance

To view an individual's chromosome inheritance, cells can be treated and photographed just prior to division. Before birth, cells can be obtained by amniocentesis, a procedure in which a physician uses a long needle to withdraw a portion of the amniotic fluid containing fetal cells. In chorionic villi sampling, cells are removed from the chorion. In any case, the fetal cells are cultured, and then a karyotype of the chromosomes is prepared (Fig. 10.2). Karyotypes can also be done using white blood cells from an adult. A karyotype displays and numbers the homologous chromosomes plus the sex chromosomes. In the male karyotype, it is possible to see that the **X chromosome** is the larger and the **Y chromosome** is the smaller of the sex chromosomes.

**Figure 10.2** **Preparation of karyotypes.**

To test the fetus for an alternation in the chromosome number or structure, fetal cells can be acquired by (*a*) amniocentesis or (*b*) chorionic villi sampling. (*c*) Karyotyping then will reveal chromosomal mutations. In this case, the karyotype shows that the newborn will have Down syndrome.

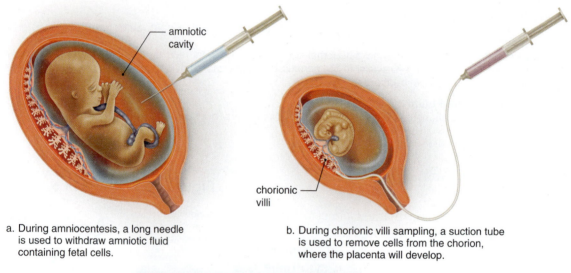

a. During amniocentesis, a long needle is used to withdraw amniotic fluid containing fetal cells.

b. During chorionic villi sampling, a suction tube is used to remove cells from the chorion, where the placenta will develop.

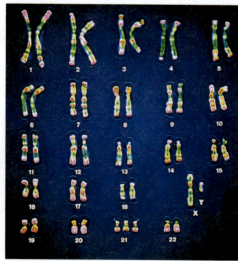

c. Karyotype of a person with Down syndrome. Note the three number 21 chromosomes.

## Numerical Chromosome Abnormalities

Gamete formation in humans involves meiosis, the type of cell division that reduces the chromosome number by one-half because the homologues separate during meiosis I. When homologues fail to separate during meiosis I, called **nondisjunction,** gametes with too few (n − 1) or too many (n + 1) chromosomes result. Nondisjunction can also occur during meiosis II if the chromatids fail to separate and the daughter chromosomes go into the same daughter cell.

### Down Syndrome (Trisomy 21)

Down syndrome, the most common autosomal trisomy in humans, is due to the inheritance of three number 21 chromosomes (Fig. 10.2c). Therefore, the syndrome is also called trisomy 21. Usually, nondisjunction occurs during oogenesis, and the egg contains two instead of one chromosome 21.

A person with Down syndrome is short, has an eyelid fold, a flat face, stubby fingers, a large fissured tongue and, unfortunately, mental retardation, which can sometimes be severe.

**Figure 10.3** **Nondisjunction of sex chromosomes.**
*a.* Nondisjunction during oogenesis produces the types of eggs shown. Fertilization with normal sperm results in the syndromes noted. (Nonviable means existence is not possible.) *b.* Nondisjunction during meiosis I of spermatogenesis results in the types of sperm shown. Fusion with normal eggs results in the syndromes noted. Nondisjunction during meiosis II of spermatogenesis results in the types of sperm shown. Fusion with normal eggs results in the syndromes shown.

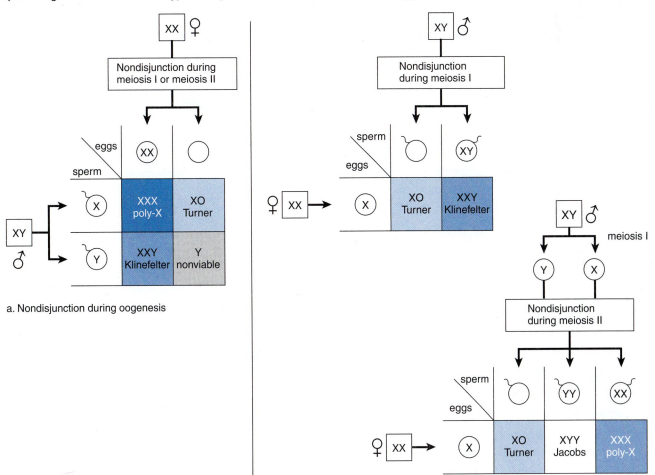

a. Nondisjunction during oogenesis

b. Nondisjunction during spermatogenesis

## Sex Chromosome Abnormalities

Figure 10.3*a* shows nondisjunction of the X chromosomes in humans during oogenesis. Three viable abnormal chromosomal types can occur.

A female with **Turner syndrome** (XO) has only one sex chromosome, an X chromosome; the O signifies the absence of the second sex chromosome. Because the ovaries never become functional, these females do not undergo puberty or menstruation, and their breasts do not develop. Generally, females with Turner syndrome have a short build, folds of skin on the back of the neck, difficulty recognizing various spatial patterns, and normal intelligence. With hormone supplements, they can lead fairly normal lives.

When an egg having two X chromosomes is fertilized by an X-bearing sperm, an individual with **poly-X syndrome** results. The body cells have three X chromosomes, and therefore 47 chromosomes. It might be supposed that poly-X females are especially feminine, but this is not the case. Although they tend to have learning disabilities, poly-X females have no apparent physical abnormalities, and many are fertile and have children with a normal chromosome count.

When an egg having two X chromosomes is fertilized by a Y-bearing sperm, a male with **Klinefelter syndrome** results. This individual is male in general appearance, but the testes are underdeveloped, and the breasts may be enlarged. The limbs of XXY males tend to be longer than average, muscular development is poor, body hair is sparse, and many XXY males have learning disabilities.

**Jacobs syndrome** can be due to nondisjunction during meiosis II of spermatogenesis. These males are usually taller than average, suffer from persistent acne, and tend to have speech and reading problems. At one time, it was suggested that XYY males were likely to be criminally aggressive, but the incidence of such behavior has been shown to be no greater than that among XY males.

Complete Table 10.1 to show how a physician would recognize each of these syndromes from a karyotype.

| Table 10.1 | Numerical Sex Chromosome Abnormalities |
|---|---|
| **Syndrome** | **Karyotype** |
| Turner | |
| Poly-X | |
| Klinefelter | |
| Jacobs | |

## Experimental Procedure: Nondisjunction*

### Building the Chromosomes

1. Obtain the following materials: 36 red pop beads, 26 blue (or green) pop beads, and eight magnetic centromeres.
2. Build four duplicated sex chromosomes (3 Xs and 1 Y) as follows:

   Two red X chromosomes: Each chromatid will have nine red pop beads. Place the centromeres so that two beads are above each centromere and seven beads are below each centromere. Bring the centromeres together.

   One blue X chromosome: Each chromatid will have nine blue pop beads. Place the centromere so that two beads are above each centromere and seven beads are below each centromere. Bring the centromeres together.

*Exercise courtesy of Victoria Finnerty, Sue Jinks-Robertson, and Gregg Orloff of Emory University, Atlanta, GA.

One blue Y chromosome: Each chromatid will have four blue pop beads. Place the centromeres so that two beads are above each centromere and two beads are below each centromere. Bring the centromeres together.

## Simulating Normal Oogenesis

Construct a primary oocyte by placing one blue and one red X chromosome together in the middle of your work area. (The blue X chromosome came from the father, and the red X chromosome came from the mother.) Have the chromosomes go through meiosis I and meiosis II. What is the chromosome constitution of each of the four meiotic products? Each "egg" has _____ (number) _____ (type) chromosome(s).

## Simulating Normal Spermatogenesis

Construct a primary spermatocyte by placing a red X and a blue Y chromosome together in the middle of your work area. (The red X chromosome came from the mother, and the blue Y chromosome came from the father.) Have the chromosomes go through meiosis I and meiosis II. What is the chromosome constitution of each of the four meiotic products? Two sperm have one _____ chromosome, and two sperm have one _____ chromosome.

## Simulating Fertilization

In the Punnett square provided here, fill in the products of fertilization using the *type* of gamete that resulted from normal oogenesis and the *types* of gametes that resulted from normal spermatogenesis. (Disregard the color of the chromosomes.)

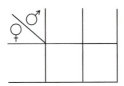

## Simulating Nondisjunction During Meiosis I

1. Construct a primary oocyte and a primary spermatocyte as before. Assume that nondisjunction occurs during meiosis I, but the chromatids separate at the centromere during meiosis II. What is the chromosome constitution of each of the four meiotic products:

   for spermatogenesis? ⬭⬭⬭⬭    for oogenesis? ○○○○

   Further note that each egg having chromosomes has one _____ (color) chromosome and one _____ (color) chromosome.

2. In the space provided here, draw a Punnett square, and fill in the products of fertilization using (a) normal sperm × *types* of abnormal eggs as in Figure 10.3*a* and (b) normal egg × *types* of abnormal sperm, as in Figure 10.3*b*. (Disregard the color of the chromosomes.)

a.                                    b.

## Conclusions

- What syndromes are the result of (a)? _____
- Are all offspring viable (capable of living)? _____ Explain. _____

_____

- What syndromes are the result of (b)? _____
- Are all offspring viable? _____ Explain. _____

_____

### Simulating Nondisjunction During Meiosis II

1. Construct a primary oocyte and a primary spermatocyte as before. Assume that meiosis I is normal, but the chromatids of the chromosomes fail to separate during meiosis II. What is the chromosome constitution of each of the four meiotic products:

for spermatogenesis?   for oogenesis?

2. In the space provided here, draw a Punnett square, and fill in the products of fertilization using (a) normal sperm × *types* of abnormal eggs, as in Figure 10.3*a*, and (b) normal egg × *types* of abnormal sperm, as in Figure 10.3*b*.

a.

b.

## Conclusions

- What syndromes are the result of (a)? _____
- Are all offspring viable? _____ Explain. _____

_____

- What syndromes are the result of (b)? _____
- Are all offspring viable? _____ Explain. _____

_____

# 10.2 Genetic Disorders: The Present

Today, genetic counselors can decide the pattern of inheritance by studying pedigrees.

## Pedigrees

A **pedigree** shows the inheritance of a genetic disorder within a family and can help determine whether any particular individual has an allele for that disorder. Then a Punnett square can be done to determine the chances of a couple producing an affected child.

In a pedigree, Roman numerals indicate the generation, and Arabic numerals indicate particular individuals in that generation. The symbols used to indicate normal and affected males and females, reproductive partners, and siblings are shown in Figure 10.4.

**Figure 10.4** Pedigree symbols.

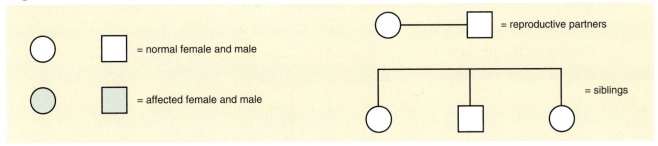

*Pedigree Analyses*

For each of the following pedigrees, determine how a genetic disorder is passed. Use Table 10.2 to help with this determination. Is the inheritance pattern autosomal dominant, autosomal recessive, or X-linked recessive? Also, decide the genotype of particular individuals in the pedigree. Remember that the *genotype* indicates the dominant and recessive alleles present and the *phenotype* is the actual physical appearance of the trait in the individual. A pedigree indicates the phenotype, and you can reason out the genotype.

| Table 10.2 | Pedigree Solution Chart | | |
|---|---|---|---|
| **Inheritance Pattern** | **Notes** | **Clues** | **Possible Genotypes** |
| Autosomal dominant (any chromosome except X or Y) | If at least one chromosome has the allele, the individual will be affected. | One or both parents are affected. Many of the children are affected. | $AA$ or $Aa$ = affected $aa$ = normal |
| Autosomal recessive (any chromosome except X or Y) | Both chromosomes must have the recessive allele for the individual to be affected. | Neither parent is affected. Few of the children are affected. | $AA$ or $Aa$ = normal; $Aa$ = carrier* $aa$ = affected |
| X-linked recessive (only the X chromosome) | The trait is only carried on the X chromosome. There must be a recessive allele on the X chromosome for the trait to be expressed. | Trait is primarily found in males. It is often passed from grandfather to grandson. | $X^A X^A$ and $X^A X^a$ = normal female; $X^A X^a$ = carrier female* $X^A Y$ = normal male $X^a X^a$ = affected female $X^a Y$ = affected male |

*A carrier is one who does not show the trait but has the ability to pass it on to his or her offspring.

1. Study the pedigree at the top of page 112:

   a. What is the inheritance pattern for this genetic disorder? _____

   b. What is the genotype of the following individuals? Use $A$ for the dominant allele and $a$ for the recessive allele.

      Generation I, individual 1: _____

      Generation II, individual 1: _____

      Generation III, individual 8: _____

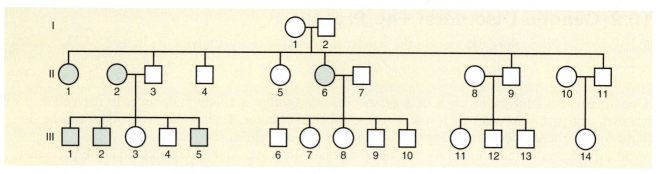

2. Study the following pedigree:

   a. What is the inheritance pattern for this genetic disorder? _____

   b. What is the genotype of the following individuals?

      Generation I, individual 1: _____

      Generation II, individual 8: _____

      Generation III, individual 1: _____

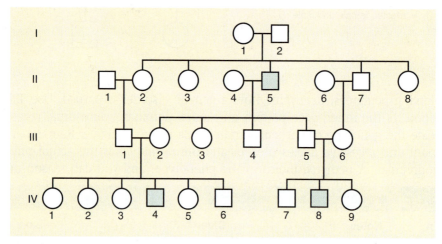

3. Study the following pedigree:

   a. What is the inheritance pattern for this genetic disorder? _____

   b. What is the genotype of the following individuals?

      Generation I, individual 1: _____

      Generation II, individual 7: _____

      Generation III, individual 4: _____

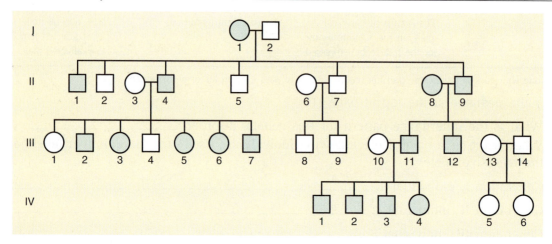

## Inheritance of a Genetic Disorder

After a counselor has decided the inheritance pattern of genetic disorder, a couple can be advised about the various chances of having a child with a disorder.

### Autosomal Dominant and Autosomal Recessive

With regard to disorders on the autosomal chromosomes, counselors are familiar with the crosses shown in Figures 10.5 and 10.6.

**Figure 10.5** Heterozygous-by-heterozygous cross.

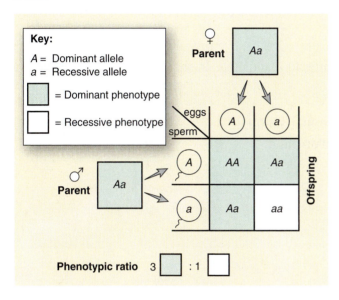

**Figure 10.6** Heterozygous-by-homozygous recessive cross.

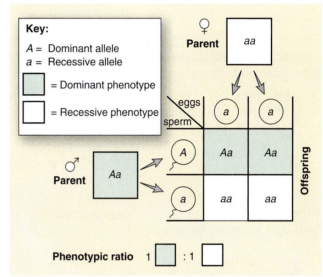

1. With reference to Figure 10.5, if a genetic disorder is recessive and the parents are heterozygous, what are the chances that an offspring will have the disorder? _____

2. With reference to Figure 10.5, if a genetic disorder is dominant and the parents are heterozygous, what are the chances that an offspring will have the disorder? _____

3. With reference to Figure 10.6, if the parents are heterozygous by homozygous recessive, and the genetic disorder is recessive, what are the chances that the offspring will have the disorder? _____

4. With reference to Figure 10.6, if the parents are heterozygous by homozygous recessive, and the genetic disorder is dominant, what are the chances that an offspring will have the disorder? _____

## Genetics Problems

1. Neurofibromatosis is a dominant disorder. If a woman is heterozygous and reproduces with a homozygous normal man, what are the chances that a child will have neurofibromatosis?

   _____

2. Cystic fibrosis is a recessive disorder. A man and a woman are both carriers for cystic fibrosis. What are the chances a child will have cystic fibrosis? _____

3. Huntington disease is a dominant disorder. Mary is 25 years old and as yet has no signs of Huntington disease. However, her mother is heterozygous for Huntington disease, but her father is free of the disorder. What are the chances that Mary will develop Huntington disease? _____

4. Phenylketonuria (PKU) is a recessive disease. Mr. and Mrs. Smith appear to be normal, but they have a child with PKU. What are the genotypes of Mr. and Mrs. Smith? _____

## X-Linked Recessive

Counselors are familiar with the crosses shown in Figures 10.7 and 10.8. In these crosses the possible genotypes and phenotypes are as follows:

**Females**
$X^B X^B$ = normal vision
$X^B X^b$ = normal vision (carrier)
$X^b X^b$ = color blind

**Males**
$X^B Y$ = normal vision
$X^b Y$ = color blind

**Figure 10.7  Cross of a normal-visioned man and a carrier woman.**

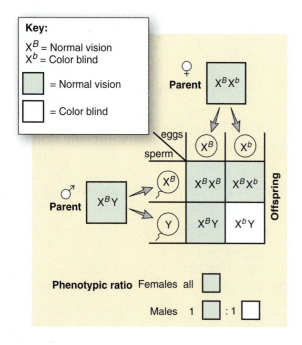

Key:
$X^B$ = Normal vision
$X^b$ = Color blind
[shaded] = Normal vision
[white] = Color blind

1. With reference to Figure 10.7, if the mother is a carrier and the father has normal vision, what are the chances that a daughter will be color blind? _____

   A daughter will be a carrier?

   _____

   A son will be color blind?

   _____

## Figure 10.8 Cross of a color-blind man and a normal woman.

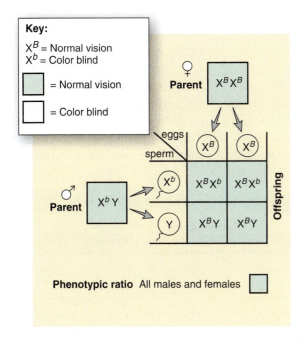

**Key:**
$X^B$ = Normal vision
$X^b$ = Color blind

☐ (shaded) = Normal vision
☐ = Color blind

Parent ♀ $X^B X^B$

eggs $X^B$ $X^B$
sperm

Parent ♂ $X^b Y$

$X^b$ → | $X^B X^b$ | $X^B X^b$ |
Y → | $X^B Y$ | $X^B Y$ |

Offspring

**Phenotypic ratio** All males and females ☐

2. With reference to Figure 10.8, if the mother has normal vision and the father is color blind, what are the chances that a daughter will be color blind? _____

A daughter will be a carrier?

_____

A son will be color blind?

_____

## Genetics Problems

1. A woman with normal color vision, whose father was color blind, marries a man with normal color vision. What do you expect to see among their offspring?

   What would you expect if it was the normal-visioned man's father who was color blind?

   _____

   _____

2. John's father is color blind but his mother is not color blind. Could John be color blind? _____ Why? _____ Is John necessarily color blind? _____ Why? _____

3. A person with Turner syndrome has hemophilia (see Table 10.3). Her mother does not have hemophilia, but her father does. In which parent did nondisjunction occur, considering that the single X came from the father? _____ Is it possible to tell if nondisjunction occurred during meiosis I or meiosis II? _____ Explain. _____

   _____

   (*Hint:* Use Table 10.2 and Figure 10.3 to help solve this problem.)

## Testing for Genetic Disorders

Tests are now available for a large number of genetic disorders (Table 10.3). For example, chromosomal tests are available for cystic fibrosis, neurofibromatosis, and Huntington disease. Blood tests can identify carriers of thalassemia and sickle-cell disease. By measuring enzyme levels in blood, tears, or skin cells, carriers of enzyme defects can also be identified for certain inborn metabolic errors, such as Tay-Sachs disease. From this information and the knowledge of the pattern of inheritance, a counselor can sometimes predict the chances of a couple having a child with the disorder.

If a woman is already pregnant, chorionic villi sampling can be done early, and amniocentesis can be done later in the pregnancy. These procedures, illustrated in Figure 10.3, allow the testing of embryonic and fetal cells, respectively, to determine if the unborn child has a genetic disorder. If so, treatment may be available even before birth, or parents may decide whether to end the pregnancy.

| Table 10.3 | Tests and Treatments for Some Human Genetic Disorders | | | |
|---|---|---|---|---|
| Name | Description | Chromosome | Incidence Among Newborns in the U.S. | Status |
| *Autosomal Recessive Disorders* | | | | |
| Cystic fibrosis | Mucus in the lungs and digestive tract is thick and viscous, making breathing and digestion difficult. | 7 | One in 2,500 Caucasians | Allele located; chromosome test now available*; treatment being investigated |
| Tay-Sachs disease | Neurological impairment and psychomotor difficulties develop early, followed by blindness and uncontrollable seizures; death usually occurs before age 5. | 15 | One in 3,600 Jews of eastern European descent | Biochemical test now available* |
| Phenylketonuria | The inability to metabolize phenylalanine; if a special diet is not begun, mental retardation develops. | 12 | One in 5,000 Caucasians | Biochemical test now available; treatment available |
| Sickle-cell disease | Sickle-shaped red blood cells cause poor circulation, anemia, and internal hemorrhaging. | 11 | One in 100 African Americans | Chromosome test now available* |
| *Autosomal Dominant Disorders* | | | | |
| Neurofibromatosis | Benign tumors occur under the skin or deeper. | 17 | One in 3,000 | Allele located; chromosome test now available* |
| Huntington disease | Minor disturbances in balance and coordination develop in middle age and progress toward severe neurological disturbances, leading to death. | 4 | One in 20,000 | Allele located; chromosome test now available* |
| *X-Linked Recessive Disorders* | | | | |
| Hemophilia A | Propensity for bleeding, often internally, due to the lack of a blood clotting factor. | X | One in 15,000 male births | Treatment available |
| Duchenne muscular dystrophy | Muscle weakness develops early and progressively intensifies until death occurs, usually before age 20. | X | One in 5,000 male births | Allele located; biochemical tests of muscle tissue available; treatment being investigated |

*Prenatal testing is done.

If a woman is not yet pregnant, how might a couple who test positive for an abnormal allele ensure that a child born to them will be normal? In most instances, it is possible to perform *in vitro fertilization* and then preimplantation genetic diagnosis (PGD) to determine the genotype of the embryo with regard to particular genetic disorders. During preimplantation analysis, one of the cells of an eight-celled embryo is removed, and its genotype is determined (Fig. 10.9). The other seven cells will continue to develop normally. Only embryos that test normal are implanted in the uterus of the female.

So far, about 500 children free of genetic disorders that run in their families have been born worldwide following PGD. In the future, PGD might be coupled with gene therapy so that any embryo would be suitable for implantation.

**Figure 10.9** **Preimplantation genetic diagnosis.**
One cell from an eight-celled embryo can be tested for abnormal alleles; if all tested alleles are normal, the seven-celled pre-embryo can complete normal development.

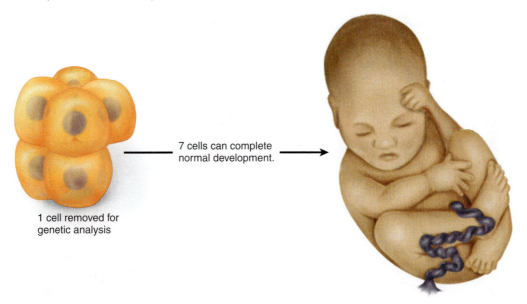

1 cell removed for genetic analysis

7 cells can complete normal development.

## 10.3 Genetic Disorders: The Future

The base sequence of the DNA in all the chromosomes is an organism's genome. Now that the Human Genome Project is finished, we know the normal order of all the 3.6 billion nucleotide bases in the human genome. Someday it will be possible to sequence anyone's genome within a relatively short time, and thereby determine what particular alterations in base sequence signify that he or she have a disorder or will have one in the future. In this laboratory, you will study how an alteration in base sequence causes a person to have sickle-cell disease.

In persons with sickle-cell disease, the red blood cells aren't biconcave disks, as are normal red blood cells—they are sickle-shaped. Sickle-shaped cells can't pass along narrow capillary passageways. They clog the vessels and break down, causing the person to suffer from poor circulation, anemia, and poor resistance to infection. Internal hemorrhaging leads to further complications, such as jaundice, episodic pain in the abdomen and joints, and damage to internal organs.

Sickle-shaped red blood cells are caused by an abnormal hemoglobin ($Hb^S$). Individuals with the $Hb^A Hb^A$ genotype are normal; those with the $Hb^S Hb^S$ genotype have sickle-cell disease and those with the $Hb^A Hb^S$ have sickle-cell trait. Persons with sickle-cell trait do not usually have sickle-shaped cells unless they experience dehydration or mild oxygen deprivation.

## Genomic Sequence for Sickle-Cell Disease

Examine Figure 10.10a and b, which show the DNA base sequence, the mRNA codons, and the amino acid sequence for a portion of the gene for $Hb^A$ and the same portion for $Hb^S$.

**Figure 10.10  Sickle-cell disease.**
Sickle-cell disease occurs when (a) the DNA base sequence in one location has changed from CTC (b) to CAC in both alleles for $Hb^A$.

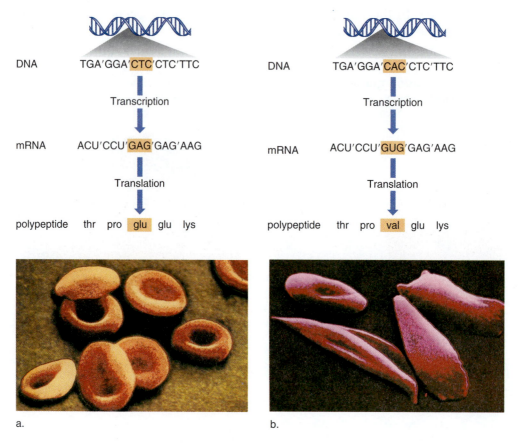

1. In what one base does $Hb^A$ differ from $Hb^S$?    $Hb^A$ _____  $Hb^S$ _____

2. What are the codons that contain this base?    $Hb^A$ _____  $Hb^S$ _____

3. What is the amino acid difference?    $Hb^A$ _____  $Hb^S$ _____

This amino acid difference causes the polypeptide chain in sickle-cell hemoglobin to pile up as firm rods that push against the plasma membrane and deform the red blood cell into a sickle shape:

## Gel Electrophoresis

If so instructed, you will carry out gel electrophoresis as directed by a kit and described in Laboratory 9, Figure 9.10. Gel electrophoresis allows us to detect the molecular difference between hemoglobin in a person with sickle-cell disease, a normal person, and a person with sickle-cell trait (Fig. 10.11). If you are *not* doing the gel electrophoresis, continue with Analyzing the Electrophoresed Gel below.

### Experimental Procedure: Gel Electrophoresis

> **Caution:** **Gel electrophoresis** Students should wear personal protective equipment: safety goggles and smocks or aprons while loading gels and during electrophoresis and protective gloves while staining.

Obtain three samples of hemoglobin provided by your kit. They are labeled sample A, B, and C. Then, as directed by your kit, carry out electrophoresis of these samples.

### Analyzing the Electrophoresed Gel

1. Sickle-cell hemoglobin ($Hb^S$) migrates slower toward the positive pole than normal hemoglobin ($Hb^A$) because the amino acid valine has no polar *R* groups, whereas the amino acid glutamate does have a polar *R* group.

2. In Figure 10.11, which lane contains only $Hb^S$, signifying that the individual is $Hb^SHb^S$? _____

3. Which lane contains only $Hb^A$, signifying that the individual is $Hb^AHb^A$? _____

4. Which lane contains both $Hb^S$ and $Hb^A$, signifying that the individual is $Hb^AHb^S$? _____

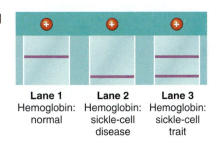

**Figure 10.11** Gel electrophoresis of hemoglobins.

Lane 1
Hemoglobin: normal

Lane 2
Hemoglobin: sickle-cell disease

Lane 3
Hemoglobin: sickle-cell trait

## Conclusion

- You are a genetic counselor of the future. A young couple seeks your advice because sickle-cell disease occurs among the family members of each. You order DNA base sequencing to be done. The results come back that at one of the loci for normal hemoglobin, each has the abnormal sequence CAC instead of CTC. The other locus is normal. What are the chances that this couple will have a child with sickle-cell disease? _____
- How is it possible for this couple to ensure that only normal embryos come to term?

_____

_____  1. What term refers to paired chromosomes arranged by size and shape?

_____  2. What pair of chromosomes is not homologous in a normal male karyotype?

_____  3. What syndrome is inherited when an egg carrying two X chromosomes is fertilized by a sperm carrying one Y chromosome?

_____  4. What abnormal meiotic event leads to the syndrome described in question 3? In which parent?

_____  5. What two types of sperm result if nondisjunction of sex chromosomes occurs during meiosis I of spermatogenesis?

_____  6. Name a common autosomal trisomy.

_____  7. What does a genetic counselor construct to show the inheritance pattern of a genetic disorder within a family?

_____  8. If an individual exhibits the dominant trait, what two genotypes are possible? (Use A for the dominant allele and a for the recessive allele.)

_____  9. What is the genotype of a man who has the sickle-cell trait?

_____ 10. The alleles of which parent, regardless of the phenotype, determine color blindness in a son?

_____ 11. If only males are affected in a pedigree, what is the likely inheritance pattern for the trait?

_____ 12. If the parents are not affected and a child is affected, what is the inheritance pattern?

## Thought Questions

13. An egg contains two X chromosomes that carry the same alleles. Did nondisjunction take place during meiosis I or meiosis II? Explain.

14. What inheritance pattern in a pedigree would allow you to decide that a trait is X-linked?

# 11

# Evidences of Evolution

## Learning Objectives

**11.1 Fossil Record**
- Define the term "fossil," and use fossils to trace the history of life.
- Understand how fossils give evidence of common descent.

**11.2 Comparative Anatomy**
- Explain how comparative anatomy gives evidence of common descent.
- Compare the human skeleton with the chimpanzee skeleton, and explain the differences on the basis of adaptation to different ways of life.

**11.3 Biochemical Evidence**
- Explain how biochemistry aids the study of the evolutionary relationships among organisms.
- Explain how biochemistry gives evidence of common descent.

## Introduction

**Evolution** is the process by which life has changed through time. A **species** is a group of similarly constructed organisms that share common genes, and a **population** is all the members of a species living in a particular area. When new variations arise that allow certain members of a population to capture more resources, these individuals tend to survive and to have more offspring than the other, unchanged members. Therefore, each successive generation will include more members with the new variation. Eventually, most members of a population and then the species will have the same **adaptations,** structures, physiology, and behavior that make an organism suited to its environment.

Adaptations to various ways of life explain why life is so diverse. However, evolution, which has been ongoing since the origin of life, is also an explanation for the unity of life. All organisms share the same characteristics of life because they can trace their ancestry to the first cell or cells. Many different lines of evidence support this hypothesis of common descent, and the more varied the evidence supporting the hypothesis, the more certain the hypothesis becomes.

In this laboratory, you will study three types of data that support the hypothesis of common descent: (1) the fossil record, (2) comparative anatomy (embryological and adult), and (3) biochemical comparison. **Fossils** are the remains or evidence of some organism that lived long ago. Fossils can be used to trace the history of life on Earth. A comparative study of the anatomy of modern groups of organisms has shown that each group has structures of similar construction called **homologous structures.** For example, all vertebrate animals have essentially the same type of skeleton. Homologous structures signify relatedness through evolution. Living organisms use the same basic molecules including ATP, the carrier of energy in cells; DNA, which makes up genes; and proteins, such as enzymes and antibodies. In this laboratory, you will analyze the activity of enzymes to determine evolutionary relationships among organisms.

| Table 11.1 | The Geological Time Scale: Major Divisions of Geological Time with Some of the Major Evolutionary Events of Each Geological Period |
|---|---|

| Era | Period | Epoch | MYA | Plant and Animal Life | |
|---|---|---|---|---|---|
| Cenozoic* | Neogene | Holocene | 0–0.01 | AGE OF HUMAN CIVILIZATION; Destruction of tropical rain forests accelerates extinctions. | |
| | | | | SIGNIFICANT MAMMALIAN EXTINCTION | |
| | | Pleistocene | 0.01–2 | Modern humans appear; modern plants spread and diversify. | |
| | | Pliocene | 2–6 | First hominids appear; modern angiosperms flourish. | |
| | | Miocene | 6–24 | Apelike mammals, grazing mammals, and insects flourish; grasslands spread; and forests contract. | |
| | Paleogene | Oligocene | 24–37 | Monkeylike primates appear; modern angiosperms appear. | |
| | | Eocene | 37–58 | All modern orders of mammals are present; subtropical forests flourish. | |
| | | Paleocene | 58–65 | Primates, herbivores, carnivores, insectivores are present; angiosperms diversify. | |
| | | | | MASS EXTINCTION: DINOSAURS AND MOST REPTILES | |
| Mesozoic | Cretaceous | | 65–144 | Placental mammals and modern insects appear; angiosperms spread and conifers persist. | |
| | Jurassic | | 144–208 | Dinosaurs flourish; birds and angiosperms appear. | |
| | | | | MASS EXTINCTION | |
| | Triassic | | 208–250 | First mammals and dinosaurs appear; forests of conifers and cycads dominate land; corals and molluscs dominate seas. | |
| | | | | MASS EXTINCTION | |
| Paleozoic | Permian | | 250–286 | Reptiles diversify; amphibians decline; and gymnosperms diversify. | |
| | Carboniferous | | 286–360 | Amphibians diversify; reptiles appear; and insects diversify. Age of great coal-forming forests. | |
| | | | | MASS EXTINCTION | |
| | Devonian | | 360–408 | Jawed fishes diversify; insects and amphibians appear; seedless vascular plants diversify and seed plants appear. | |
| | Silurian | | 408–438 | First jawed fishes and seedless vascular plants appear. | |
| | | | | MASS EXTINCTION | |
| | Ordovician | | 438–510 | Invertebrates spread and diversify; jawless fishes appear; nonvascular plants appear on land. | |
| | Cambrian | | 510–543 | Marine invertebrates with skeletons are dominant and invade land, and marine algae flourish. | |
| Precambrian time | | | 600 | Oldest soft-bodied invertebrate fossils | |
| | | | 1,400–700 | Protists evolve and diversify. | |
| | | | 2,000 | Oldest eukaryotic fossils | |
| | | | 2,500 | $O_2$ accumulates in atmosphere | |
| | | | 3,500 | Oldest known fossils (prokaryotes) | |
| | | | 4,500 | Earth forms. | |

* Many authorities divide the Cenozoic era into the Tertiary period (contains Paleocene, Eocene, Oligocene, Miocene, and Pliocene) and the Quaternary period (contains Pleistocene and Holocene).

# 11.1 Fossil Record

The **geological time scale** (Table 11.1) pertains to the history of the earth from its formation 4 to 4.5 billion years ago to the present. The ages of rocks can be measured in years by analyzing naturally occurring radioactive elements found in minute quantities in certain rocks and minerals. Before this method was discovered, geologists depended on a relative dating system; that is, they reasoned that any given stratum (layer of sediment) is younger than the stratum of the earth's crust just beneath it.

Because certain fossils are associated with particular strata, geologists are able to relate various strata around the world. *Fossils* are the remains and traces of past life or any other direct evidences of past life. By the 1860s, fossil-containing rocks in western Europe had been divided into three great eras: Paleozoic (ancient life), Mesozoic (middle life), and Cenozoic (recent life). Each era was divided into periods, and periods, in turn, were divided into epochs. The first period of the Paleozoic is called the Cambrian, and the time before the Cambrian period is called the Precambrian. The oldest fossils that have been found date from the Precambrian. For reasons that are still being explored, the fossil record improves dramatically starting with the Cambrian period.

Notice in Table 11.1 that divisions of the geological time scale generally tend to become increasingly shorter: For example, the Cenozoic era is much shorter than the Paleozoic era, and the Neogene period is much shorter than the Paleogene period. Notice also that time is measured as "millions of years ago," and therefore, larger numbers represent an earlier time than smaller numbers.

Use Table 11.1 to answer the following questions:

1. During the _____ era and the _____ period, the first vascular plants appeared. How many million years ago was this? _____

2. During the _____ era and the _____ period, angiosperm (flowering plant) diversity occurred. How many million years ago was this? _____

Use Figure 11.1 to answer the following questions:

3. During what period did the trilobites appear? _____ When did they become extinct? _____

4. During what period did the insects first appear? _____ How many million years ago was this? _____

Use Figure 11.2 to answer the following questions:

5. During what period were the cycads most abundant? _____ What types of animals were also prevalent at this time? _____

6. During what period did the angiosperms evolve? _____ How do you know they are the most prominent plants today? _____

_____

**Figure 11.1  Geological history of selected animals.**
The relative abundance of each group during any particular time period is indicated by the width of the band.

# Figure 11.2 Geological history of selected algae, fungi, and plants.

The relative abundance of each group during any particular time period is indicated by the width of the band.

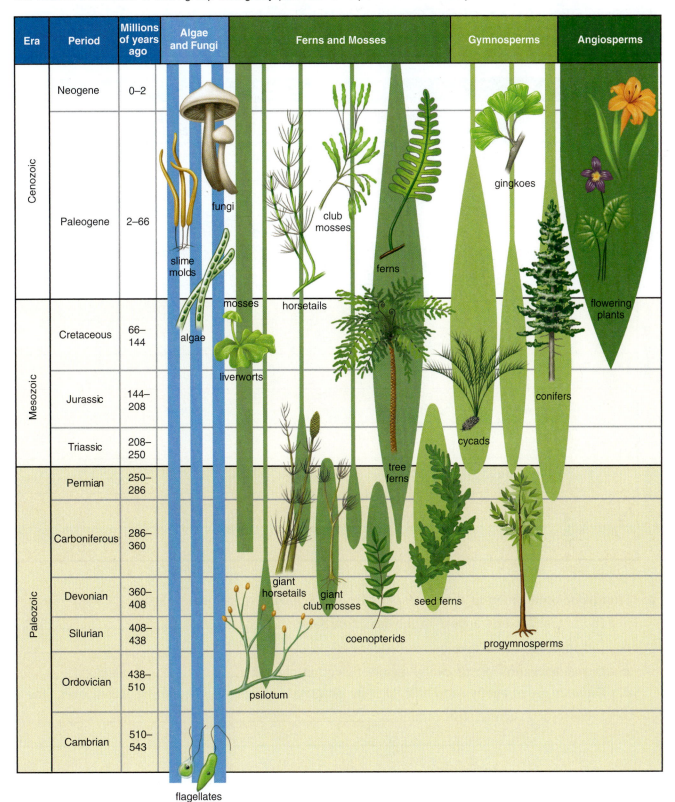

1. Obtain a box of selected fossils from each geological era (Cenozoic, Mesozoic, Paleozoic). Fill in Table 11.2 from the identification key or the fossil labels.

| Era | Period | Type of Fossil (Phylum, Class, or Common Name) | Description (May Include a Sketch) |
|---|---|---|---|
| **Table 11.2** | | **Fossils** | |
| Cenozoic | | | |
| Cenozoic | | | |
| Cenozoic | | | |
| Mesozoic | | | |
| Mesozoic | | | |
| Mesozoic | | | |
| Paleozoic | | | |
| Paleozoic | | | |
| Paleozoic | | | |

2. From your observation of the fossils, answer the following questions:

   a. Did organisms first appear in the sea or on land? _____

   b. Did invertebrates evolve before vertebrates? _____

   c. Did cone-bearing plants evolve before flowering plants? _____

   d. Do your observations show that, despite observed changes, fossils can be linked over time because of a similarity in form? _____

   _____

# 11.2 Comparative Anatomy

In the study of evolutionary relationships, organisms or parts of organisms are said to be **homologous** if they exhibit similar basic structures and embryonic origins. If these organisms or parts of organisms are similar in function only, they are said to be **analogous.** Only homologous structures indicate an evolutionary relationship and are used to classify organisms.

## Comparison of Adult Vertebrate Forelimbs

The limbs of vertebrates are homologous structures. Homologous structures share a basic pattern, although there may be specific differences. The similarity of homologous structures is explainable by descent from a common ancestor.

### Observation: Vertebrate Forelimbs

1. The central diagram in Figure 11.3 represents the forelimb bones of the ancestral vertebrate. The basic components are the humerus (h), ulna (u), radius (r), carpals (c), metacarpals (m), and phalanges (p) in the five digits.
2. Carefully compare and label in Figure 11.3 the corresponding forelimb bones of the frog, the lizard, the bird, the bat, the cat, and the human. In particular, note the specific modifications that have occurred in some of the bones to meet the demands of a particular way of life.
3. Fill in Table 11.3 to indicate which bones in each specimen appear to most resemble the ancestral condition and which most differ from the ancestral condition. Use the letter symbols from Figure 11.3.

| Table 11.3 | Comparison of Vertebrate Forelimbs | |
|---|---|---|
| **Animal** | **Bones That Resemble Common Ancestor** | **Bones That Differ from Common Ancestor** |
| Frog | | |
| Lizard | | |
| Bird | | |
| Bat | | |
| Cat | | |
| Human | | |

4. Relate the change in bone structure to mode of locomotion in two examples.

   Example 1: _____

   Example 2: _____

# Figure 11.3 Vertebrate forelimbs.

Because all vertebrates evolved from a common ancestor, their forelimbs share homologous structures.

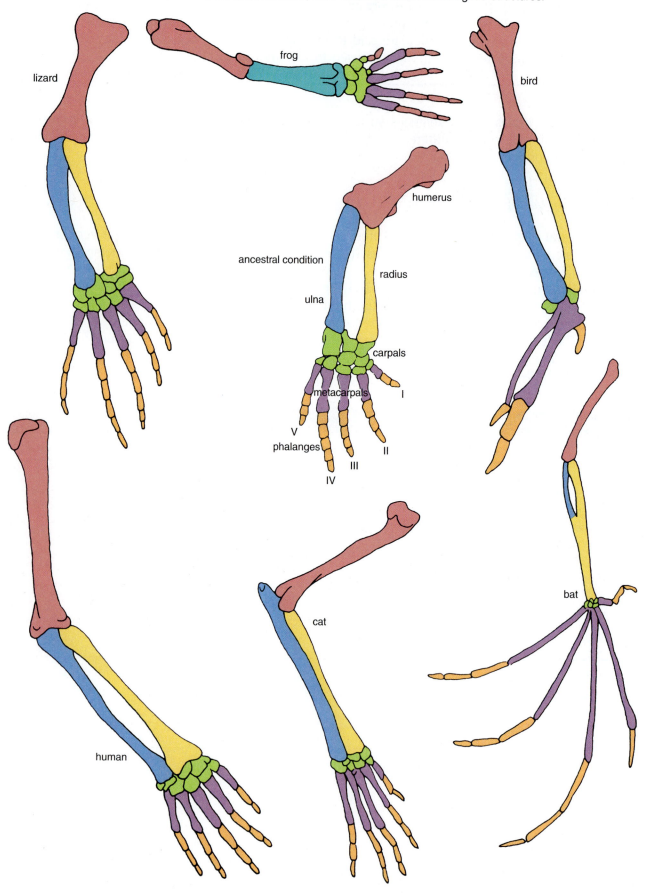

## Comparison of Chimpanzee and Human Skeletons

Chimpanzees and humans are closely related, as is apparent from the comparison of the classifications of humans and chimpanzees in Table 11.4. Are chimpanzees and humans both primates?

_____ At what category does the classification of chimpanzees and the classification of humans first differ? _____

| Table 11.4 | Comparison of Human and Chimpanzee Classifications | |
|---|---|---|
| **Classification** | **Chimpanzees** | **Humans** |
| Domain | Eukarya | Eukarya |
| Kingdom | Animalia | Animalia |
| Phylum | Chordata | Chordata |
| Class | Mammalia | Mammalia |
| Order | Primate | Primate |
| Family | Pongidae | Hominidae |
| Genus | *Pan* | *Homo* |
| Species | *troglodytes* | *sapiens* |

### Observation: Chimpanzee and Human Skeletons

**Comparison of Skeletons**

Chimpanzees are arboreal and climb in trees. While on the ground, they tend to knuckle-walk, with their hands bent. Humans are terrestrial and walk erect.

Examine chimpanzee and human skeletons (Fig. 11.4), and answer the following questions:

1. **Head and torso:** Where are the head and trunk with relation to the hips and legs—thrust forward over the hips and legs or balanced over the hips and legs? *Record your answer in Table 11.5.*
2. **Spine:** Which animal has a long and curved lumbar region, and which has a short and stiff lumbar region (top set of red arrows)? *Record your answer in Table 11.5.*

   How does this contribute to an erect posture in humans? _____

   _____

3. **Pelvis:** Chimps sway when they walk because lifting one leg throws them off balance. Which animal has a narrow and long pelvis, and which has a broad and short pelvis (bottom set of red arrows)? *Record your answer in Table 11.5.*
4. **Femur:** In humans, the femur better supports the trunk. In which animal is the femur angled between articulations with the pelvic girdle and the knee? In which animal is the femur straight with no angle? *Record your answer in Table 11.5.*
5. **Knee joint:** In humans, the knee joint is modified to support the body's weight. In which animal is the femur larger at the bottom and the tibia larger at the top? *Record your answer in Table 11.5.*
6. **Foot:** In humans, the foot is adapted for walking long distances and running with less chance of injury. In which animal is the big toe opposable? Which foot has an arch? *Record your answers in Table 11.5.*

**Figure 11.4** **Adult skeletons.**

Chimpanzee skeleton (a) compared with a human skeleton (b). See the text for an explanation of the red arrows.

(© 2001 Time Inc. reprinted by permission.)

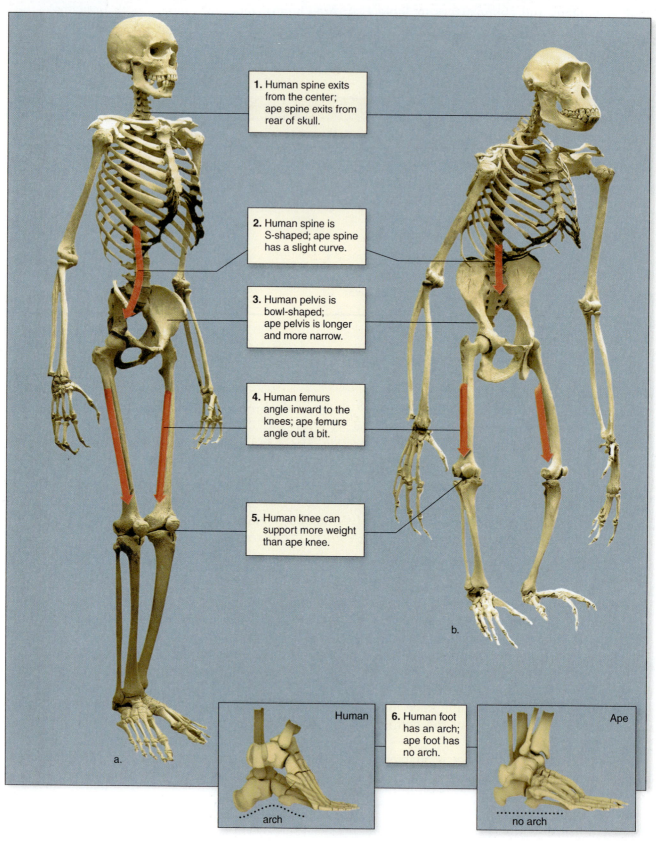

1. Human spine exits from the center; ape spine exits from rear of skull.

2. Human spine is S-shaped; ape spine has a slight curve.

3. Human pelvis is bowl-shaped; ape pelvis is longer and more narrow.

4. Human femurs angle inward to the knees; ape femurs angle out a bit.

5. Human knee can support more weight than ape knee.

6. Human foot has an arch; ape foot has no arch.

Human

arch

Ape

no arch

a.

b.

| Table 11.5 | Comparison of Chimpanzee and Human Skeletons | |
|---|---|---|
| **Skeletal Part** | **Chimpanzee** | **Human** |
| Head and torso Spine | | |
| Pelvis Femur Knee joint | | |
| Foot: Opposable toe | | |
| Arch | | |

**Figure 11.5  Chimpanzee skull compared with a human skull.**
Although the general shapes of these skulls are different, they have similarities.

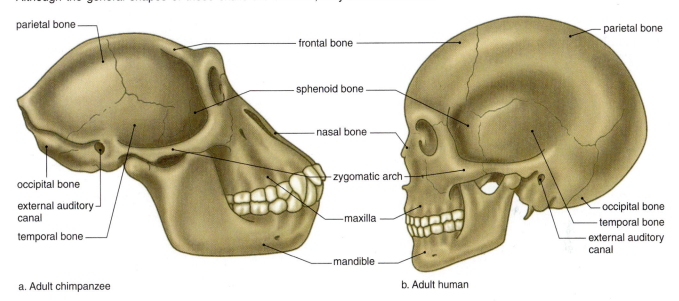

a. Adult chimpanzee

b. Adult human

7. Examine the position and shape of the parietal bones in both the chimpanzee and human skulls (Fig. 11.5). How does the chimpanzee skull differ from the human skull in this respect? _____

8. Compare the shape and position of the occipital bones in the chimpanzee and human skulls.

_____

9. How does the difference in the position of the foramen magnum, a large opening in the base of the skull for the spinal cord, correlate with the posture and stance of the two organisms (see Fig. 11.4)?

_____

10. Compare the slope of the frontal bones of the chimpanzee and human skulls. How are they different? _____

11. For which skull is the supraorbital ridge (the region of frontal bone just above the eye socket) thicker? _____

**12.** What is the position of the mouth and chin in relation to the profile for each skull? _____

_____

What effect has the evolutionary change in the positions of these bones had on the shape of the face? _____

**13.** Examine the teeth in the adult chimpanzee and adult human skulls. Are the shapes and types of teeth similar in both? _____ Diet can account for many of the observed differences. Humans are omnivorous. A diet rich in meat does not require strong grinding teeth or well-developed facial muscles. Chimpanzees are vegetarians, and a vegetarian diet requires strong facial muscles that attach to bony projections.

## Comparison of Vertebrate Embryos

The anatomy shared by vertebrates extends to their embryological development. For example, as embryos, they all have a post-anal tail, somites (segmented blocks of mesoderm lying on either side of the notochord), and paired pharyngeal pouches. In aquatic animals, these pouches become functional gills (Fig. 11.6). In humans, the first pair of pouches becomes the cavity of the middle ear and auditory tube, the second pair becomes the tonsils, and the third and fourth pairs become the thymus and parathyroid glands.

**Figure 11.6** **Vertebrate embryos.**
During early developmental stages, vertebrate embryos have certain characteristics in common.

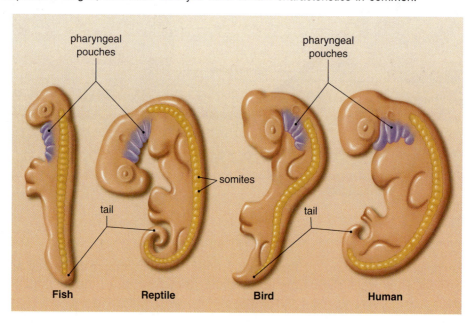

### Observation: Chick and Pig Embryos

1. Obtain prepared slides of vertebrate embryos at comparable stages of development. Observe each of the embryos using a binocular dissecting microscope.
2. List five similarities of the embryos:

   a. _____

   b. _____

   c. _____

**d.** _____

**e.** _____

**3.** Why do the embryos resemble one another so closely? _____
_____

# 11.3 Biochemical Evidence

Almost all living organisms use the same basic biochemical molecules, including DNA, ATP, and many identical and nearly identical enzymes. In addition, living organisms use the same DNA triplet code and the same 20 amino acids in their proteins. There is no obvious functional reason these elements need to be so similar. Therefore, their similarity is best explained by descent from a common ancestor.

## Protein Differences

According to the **protein clock theory,** the number of amino acid changes between organisms is proportional to the length of time since two organisms began evolving separately from a common ancestor. Why should that be? _____

The sequence of amino acids in **cytochrome c,** a carrier of electrons in the electron transport chain found in mitochondria and chloroplasts, has been determined in a variety of organisms. Figure 11.7 lists the number of differences between amino acid sequences for several of these.

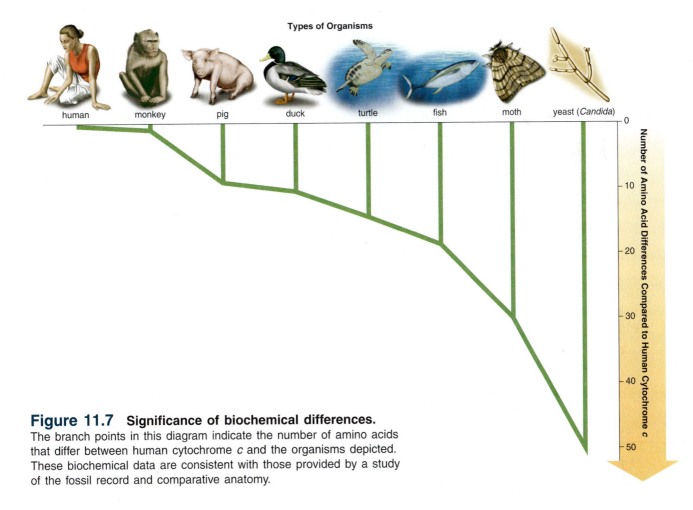

**Figure 11.7** **Significance of biochemical differences.**
The branch points in this diagram indicate the number of amino acids that differ between human cytochrome _c_ and the organisms depicted. These biochemical data are consistent with those provided by a study of the fossil record and comparative anatomy.

## Protein Similarities

The immune system makes **antibodies** (proteins) that react with foreign proteins, termed **antigens.** The antigen-antibody reaction is specific. An antibody will react only with its particular antigen. In the laboratory, this reaction can be observed when a precipitate, a substance separated from the solution, appears.

Biochemists have used the antibody-antigen reaction to determine the degree of relatedness between animals. In one technique, human serum (containing human proteins) is injected into the bloodstream of a rabbit, and the rabbit makes antibodies against the human serum. Some of the rabbit's blood is then drawn off, and the sensitized serum that contains the antibodies is separated from it. This sensitized rabbit serum will react strongly (determined by the amount of precipitate) against a new sample of human blood serum. The rabbit serum also will react against serum from other animals. The more closely related these animals are to humans, the more the precipitate forms (Fig. 11.8).

**Figure 11.8** **Antigen-antibody reaction.**
When antigens react with antibodies, a complex forms that appears as a precipitate.

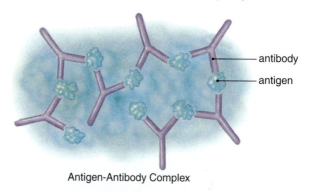

antibody
antigen

Antigen-Antibody Complex

### Experimental Procedure: Protein Similarities

1. Obtain a chemplate (a clear glass tray with wells), one bottle of synthetic *human blood serum,* one bottle of synthetic *rabbit blood serum,* and five bottles (I–V) of *blood serum test solution.*
2. Put two drops of synthetic rabbit blood serum in each of the six wells in the chemplate. Label the wells 1–6.
3. Add two drops of synthetic human blood serum to each well, and stir with the plastic stirring rod attached to the chemplate. The rabbit serum has now been "sensitized" to human serum. (This simulates the production of antibodies in the rabbit's bloodstream in response to the human blood proteins.)
4. Rinse the stirrer. (The large cavity of the chemplate may be filled with water to facilitate rinsing.)
5. Add four drops of *blood serum test solution III* (tests for human blood proteins) to well 6.

   Describe what you see. _____

   _____

   This well will serve as the basis by which to compare all the other samples of test blood serum.
6. Now add four drops of *blood serum test solution I* to well 1. Stir and observe. Rinse the stirrer. Do the same for each of the remaining *blood serum test solutions (II–V)*—adding II to well 2, III to well 3, and so on. Be sure to rinse the stirrer after each use.

7. At the end of 10 and 20 minutes, record the amount of precipitate in each of the six wells in Figure 11.9. Well 6 is recorded as having ++++ amount of precipitate after both 10 and 20 minutes. Compare the other wells with this well (+ = trace amount; 0 = none). Holding the plate slightly above your head at arm's length and looking at the underside toward an overhead light source will allow you to more clearly determine the amount of precipitate.

## Figure 11.9  Biochemical evidence of evolution.
The greater the amount of precipitate, the more closely related an animal is to humans.

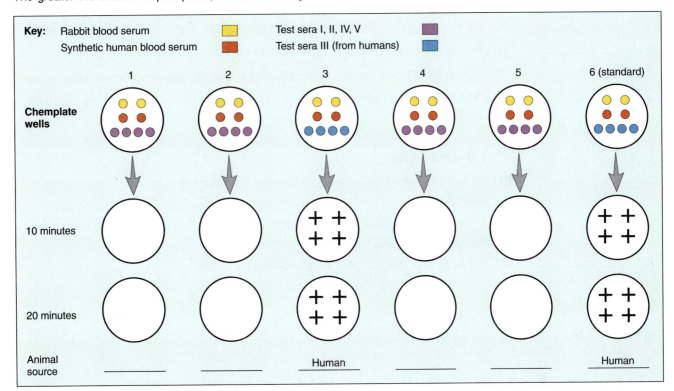

## Conclusions

- The last row in Figure 11.9 tells you that the test serum in well 3 is from a human. How do your test results confirm this? _____
- Aside from humans, the test sera (supposedly) came from a pig, a monkey, an orangutan, and a chimpanzee. Which is most closely related to humans—the pig or the chimpanzee?_____
- Judging by the amount of precipitate, complete the last row in Figure 11.9 by indicating which serum you believe came from which animal. On what do you base your conclusions?

_____

_____

_____  1. State the simplest definition of evolution.

_____  2. What does the time scale of life encompass—tens of millions or hundreds of millions of years?

_____  3. What is the best explanation for the fact that fossils do not resemble their modern-day representatives?

_____  4. Fossils are the _____ of past life.

_____  5. All vertebrates go through similar embryological stages. What does this suggest?

_____  6. Which skeleton—chimpanzee or human—has a narrow and long pelvis?

_____  7. Which has thicker supraorbital ridges, the chimpanzee skull or the human skull?

_____  8. Which term—homologous or analogous—means that components are similar in structure?

_____  9. During development, all vertebrates have _____, even though only fish have gill slits as adults.

_____  10. In this laboratory, what type of biochemical reaction was used to determine relatedness?

_____  11. What does it indicate if antibodies to the serum of one species react strongly against the serum of another species?

_____  12. All apes and humans are members of what order?

## Thought Questions

13. If a characteristic is found in bacteria, fungi, pine trees, snakes, and humans, when did it most likely evolve? Why?

14. What do mutations have to do with amino acid changes in a protein?

# 12
# Microbiology

## Learning Objectives

**12.1 Bacteria**
- Relate the structure of a bacterium to its ability to cause disease.
- Observe and describe different bacterial cell shapes.
- Practice and describe the Gram staining protocol and its usefulness.
- Identify and state the importance of cyanobacteria.

**12.2 Protists**
- Distinguish between algae on the basis of color and organization.
- Distinguish protozoans on the basis of locomotion.
- Describe the plasmodium of a plasmodial slime mold.

**12.3 Fungi**
- Recognize the reproductive structures of sac and club fungi.

## Introduction

In today's laboratory, we are studying groups of organisms—bacteria, protists, and fungi—in which at least some, if not all or many, members are microscopic. You may recall that all living things are classified as either prokaryotes or eukaryotes on the basis of whether they lack or have a nucleus. Bacteria are prokaryotes in the domain Bacteria, while protists and fungi are eukaryotes in the domain Eukarya. This means that bacteria, placed in a different domain from the other two, are distantly related to them. In this laboratory, we will study bacteria chiefly as pathogenic organisms. The protists are a diverse group because they include the photosynthetic algae, the often motile protozoans, and the somewhat funguslike slime molds. Unlike the other two groups, which are unicellular, fungi are usually multicellular. Fungi reproduce by producing windblown spores, and along with bacteria are most often saprotrophic. That means that they release digestive enzymes into the environment and absorb the products of digestion across the plasma membrane.

| | Domain | Cell Structure | Nutrition |
|---|---|---|---|
| **Bacteria** | Bacteria | Unicellular | Most heterotrophic |
| **Protists** | | | |
| Protozoans | Eukarya | Unicellular | Heterotrophic |
| Algae | Eukarya | Unicellular, colonial, filamentous, or multicellular | Photosynthetic |
| Slime molds | Eukarya | Unicellular stage and multinucleated plasmodium | Heterotrophic |
| **Fungi** | Eukarya | Multicellular | Heterotrophic |

# 12.1 Bacteria

In this laboratory you will first relate the general structure of a bacterium to its ability to cause disease. The specific shape, growth habit, and staining characteristics of bacteria are often used to identify them. Therefore, you will observe a variety of bacteria using the microscope. You will also perform one of the most important and discriminating tests, the Gram-staining protocol. Aside from their medical importance, bacteria are essential in ecosystems because, along with fungi, they are decomposers that break down dead organic remains, and thereby return inorganic nutrients to plants.

## Pathogenic Bacteria

Pathogenic bacteria are infectious agents that cause disease. Infectious bacteria are able to invade and multiply within a host. Some also produce a toxin. Antibiotic therapy is often an effective treatment against a bacterial infection.

We will explore how it is possible to relate the structure of a bacterium to its ability to be invasive and avoid destruction by the immune system. We will also consider what morphophysiological attributes allow bacteria to be resistant to antibiotics and to pass the necessary genes on to other bacteria.

### Observation: Structure of a Bacterium

1. Study the structure of a generalized bacterium in Figure 12.1, and if available, examine a model or view a CD-Rom of a bacterium.
2. Identify:

   **capsule,** a gel-like coating outside the cell wall. Capsules often allow bacteria to stick to surfaces such as teeth. They also prevent phagocytic white blood cells from taking them up and destroying them.

   **fimbriae,** hairlike bristles that allow adhesion to surfaces. This can be how a bacterium clings to and gains access to the body prior to an infection.

   **sex pilus,** elongated, hollow appendage used to transfer DNA to other cells. Genes that allow bacteria to be resistant to antibiotics can be passed in this manner.

   **flagellum,** a rotating filament that pushes the cell forward.

   **cell wall,** a structure that provides support and shapes the cell. Antibiotics that prevent the formation of a cell wall are most effective against Gram-positive rather than Gram-negative bacteria.

   **plasma membrane,** a sheet that surrounds the cytoplasm and regulates entrance and exit of molecules. Resistance to antibiotics can be due to plasma membrane alterations that do not allow the drug to bind to the membrane or cross the membrane, or to a plasma membrane that increases the elimination of the drug from the bacteria.

   **ribosomes,** site of protein synthesis. Some bacteria possess antibiotic-inactivating enzymes that make them resistant to antibiotics.

   **nucleoid,** the location of the bacterial chromosome.

## Figure 12.1  Generalized structure of a bacterium.

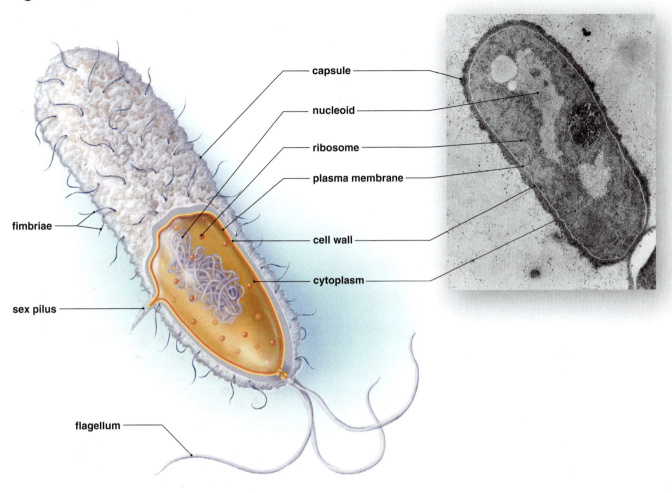

**Conclusions**

- Which portions of a bacterial cell aid the ability of a bacterium to cause infections?

  _____

- Which portions of a bacterial cell aid the ability of a bacterium to be resistant to
  antibiotics? _____

Also, some bacteria contain **plasmids,** small rings of DNA that replicate independently of the chromosomes and can be passed to other bacteria. Genes that allow bacteria to be resistant to antibiotics are often located in a plasmid.

## Identification of Bacteria by Morphology

Shape and arrangement help identify bacteria. Most bacteria have one of three shapes. Bacteria that are spherical are called cocci (sing., **coccus**)(Fig. 12.2*a*). Bacteria that are rod-shaped are called bacilli (sing., **bacillus**) (Fig. 12.2*b*). Spiral bacteria are called spirilla, (sing., **spirillum**)(Fig. 12.2*c*). The bacteria featured in Figure 12.2 are all single cells; therefore, they are unicellular.

Bacteria have a variety of shapes and arrangements in addition to the three mentioned. The cocci and bacilli can appear in a variety of sizes: large or small spheres, long or short rods. There are some intermediate forms called coccobacilli. Curved rods are called vibrios. Some bacteria exist as chains, packets, or clusters of cells. Diplococcus means that a coccus forms a chain consisting of only two bacteria. A coccus that forms a longer chain is called a *streptococcus*. A chain of bacilli is called a *streptobacillus*. Packets of four cells are called a tetrad, while packets of eight cells are called a sarcina. More common are bacteria that form clusters. When the cluster is made of cocci, the organism is a *staphylococcus*.

### Observation: Cell Morphologies

1. View the microscope slides of bacteria that are on display. What magnification is required to view bacteria?
2. Identify the three different shapes of bacteria, using Figure 12.2 as a guide.

### Figure 12.2 Shapes of bacteria.

*a.* Streptococci, which exist as chains of cocci, cause a number of illnesses, including strep throat. *b. Escherichia coli,* which lives in your intestine, is a bacillus with flagella. *c. Treponema pallidum,* the cause of syphillis, is a spirillum.

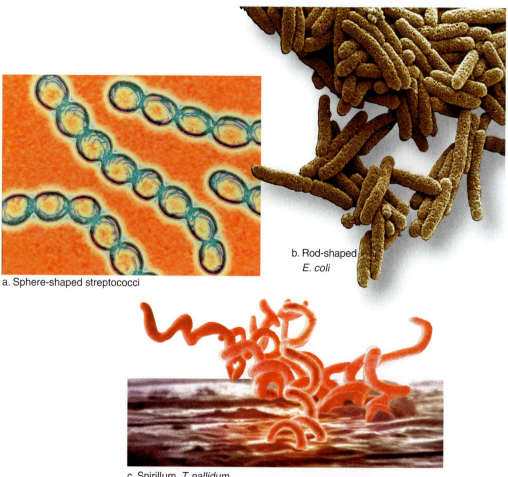

a. Sphere-shaped streptococci

b. Rod-shaped *E. coli*

c. Spirillum, *T. pallidum*

### Agar Plates

**Agar** is a semisolid medium used to grow bacteria.

1. View agar plates that have been inoculated with bacteria and then incubated. Notice the "colonies" of bacteria growing on the plates. Each colony contains cells that are all descended from one original cell.
2. Compare the colonies' color, surface, and margin, and note your observations in Table 12.1.

| Table 12.1 | Agar Staining |
|---|---|
| **Plate Number** | **Description of Colonies** |
| | |
| | |
| | |

## Identification of Bacteria by Gram Stain

Most bacterial cells are protected by a cell wall that contains a unique molecule called peptidoglycan. Bacteria are commonly differentiated by using the Gram stain procedure, which distinguishes bacteria that have a thick layer of peptidoglycan (Gram-positive) from those that have a thin layer of peptidoglycan (Gram-negative). Gram-positive bacteria retain a crystal violet-iodine complex and stain blue-purple, whereas Gram-negative bacteria decolorize and counterstain red-pink with safranin.

### Experimental Procedure: Gram Stain

1. Use one designated square of a slide that has six squares.
2. With a sterile cotton swab, obtain a sample from around your teeth or inside your nose.
3. Carefully roll the swab across your allotted square. Body samples must be spread out thinly and evenly on the slide.
4. Allow the smear to air-dry.
5. Fix the smear by flooding the slide with *absolute methanol* for 1 minute. Allow the smear to dry before staining.
6. Flood the smear with *Gram Crystal Violet*, and wait for 1 to 2 minutes.
7. Gently rinse off the crystal violet with cold tap water.
8. Flood the smear with *Gram Iodine*, and allow it to react for 1 minute.
9. Gently rinse off the iodine with cold tap water.
10. Gently rinse the smear with *Gram Decolorizer* until the solution rinses colorlessly from the slide (approximately 20 to 30 seconds).
11. Immediately rinse the smear with cold tap water.
12. Flood the smear with *Gram Safranin*, and allow it to stain for 15 to 30 seconds.
13. Gently rinse off the safranin with cold tap water.
14. Blot off excess water with a paper towel, and allow the smear to air-dry.
15. Examine microscopically. (This will require the use of the oil immersion lens.)

### Conclusion

The Gram stain is one way to distinguish bacteria from each other. Are these bacteria Gram-positive or Gram-negative? _____

# Cyanobacteria

Some bacteria are photosynthesizers that use solar energy to produce their own food. Cyanobacteria are believed to have arisen some 3.7 billion years ago and are thought to have been the first organisms to release oxygen into the atmosphere. Their importance as a source of oxygen, even today, should not be underemphasized. At one time, cyanobacteria were identified as blue-green algae, but now we know they are a type of bacterium. Keep in mind that they have isolated thylakoids and not chloroplasts.

## Observation: Cyanobacteria

### Oscillatoria

1. Prepare a wet mount of an *Oscillatoria* culture, if available, or examine a prepared slide, using high power (45×) or oil immersion (if available). This is a filamentous cyanobacterium with individual cells that resemble a stack of pennies (Fig. 12.3a).
2. *Oscillatoria* takes its name from the characteristic oscillations that you may be able to see if your sample is alive. If you have a living culture, are oscillations visible? _____

### Anabaena

1. Prepare a wet mount of an *Anabaena* culture, if available, or examine a prepared slide, using high power (45×) or oil immersion (if available). This is also a filamentous cyanobacterium, although its individual cells are barrel-shaped (Fig. 12.3b,c).
2. Note the thin nature of this strand. If you have a living culture, what is its color? _____ (Prepared slides are artificially stained and may not resemble the color of the living cyanobacterium.)

## Figure 12.3 *Oscillatoria* and *Anabaena*.

a. *Oscillatoria* is a filamentous cyanobacterium (magnification ×120). b. c. *Anabaena* is a filamentous cyanobacterium with heterocysts where atmospheric nitrogen ($N_2$) is converted to ammonia.

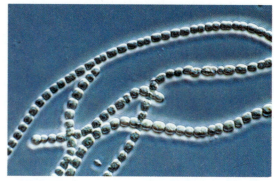

b. Photomicrograph

heterocyst

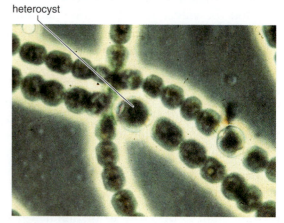

c. Photomicrograph

a. Photomicrograph

500 µm

# 12.2 Protists

Protists (kingdom Protista) are all eukaryotes, even though they may be unicellular. The protists include the algae, which photosynthesize in the same manner as plants; the protozoans, heterotrophic by ingestion; and the slime molds, which creep along the forest floor as a plasmodium.

## Algae

The algae, whether green algae, red algae, brown algae, or golden-brown algae, all photosynthesize, as do plants. Why aren't they considered plants? Because they never protect the zygote and other reproductive structures the way plants do. Aside from releasing oxygen into the environment, algae play an important role in aquatic ecosystems—both freshwater and marine—because they are producers. Producers produce food for themselves and all members of an ecosystem.

### Observation: A Sampling of Algae

To exemplify algae, you will examine a filamentous form (*Spirogyra*), a colonial form (*Volvox*), and a unicellular form (diatoms). The seaweeds seen along the coasts are also algae.

1.  Obtain and examine a slide of *Spirogyra* (Fig. 12.4). The most prominent feature of the cells is the spiral, ribbonlike chloroplast. The nucleus is in the center of the cell, anchored by cytoplasmic strands. Your slide may show **conjugation,** a sexual means of reproduction illustrated in Figure 12.4b. If it does not, obtain a slide that does show this process. Conjugation tubes form between two adjacent filaments, and the contents of one set of cells enter the other set. As the nuclei fuse, a zygote is formed. The zygote overwinters, and in the spring, meiosis and subsequently germination occur.

**Figure 12.4  *Spirogyra.***
*a. Spirogyra,* a filamentous green alga, in which each cell has a ribbonlike chloroplast. *b.* During conjugation, the cell contents of one filament enter the cells of another filament. Zygote formation follows.

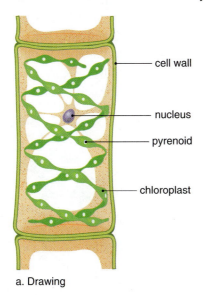

cell wall

nucleus

pyrenoid

chloroplast

a. Drawing

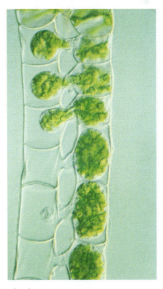

20 μm

b. Conjugation

2. Obtain and examine a slide of *Volvox* (Fig. 12.5). *Volvox* is a green algal colony. It is motile (capable of locomotion) because the thousands of cells that make up the colony have flagella. These cells are connected by delicate cytoplasmic extensions.

   *Volvox* is capable of both asexual and sexual reproduction. Certain cells of the adult colony can divide to produce **daughter colonies** (Fig. 12.5c) that reside for a time within the parental colony. A daughter colony escapes the parental colony by releasing an enzyme that dissolves away a portion of the matrix of the parental colony. During sexual reproduction, some colonies of *Volvox* have cells that produce sperm, and others have cells that produce eggs.

**Figure 12.5** *Volvox.*

(*a*) The adult colony contains (*b*) many individual cells and produces (*c*) daughter colonies.

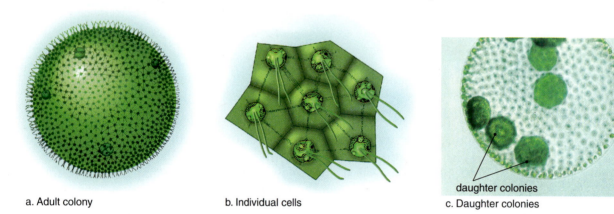

a. Adult colony      b. Individual cells      c. Daughter colonies

daughter colonies

200 µm

3. Obtain and examine a slide of diatoms (Fig. 12.6). Diatoms possess a yellow-brown pigment in addition to chlorophyll (see Fig. 12.6). The diatom cell wall is in two sections, with the larger one fitting over the smaller as a lid fits over a box. Since the cell wall is impregnated with silica, diatoms are said to "live in glass houses." The glass cell walls of diatoms do not decompose, so they accumulate in thick layers subsequently mined as diatomaceous earth and used in filters and as a natural insecticide. Diatoms, being photosynthetic and extremely abundant, are important food sources for the small heterotrophs (organisms that must acquire food from external sources) in both marine and freshwater environments.

**Figure 12.6** **Diatom shells.**

The overlapping shells of a diatom are impregnated with silica. Scientists use their delicate markings to identify the particular species.

## Protozoans

The term protozoan refers to unicellular eukaryotes and is often restricted to unicellular, heterotrophic eukaryotes that ingest food by forming **food vacuoles.** Other vacuoles, such as **contractile vacuoles** that rid the cell of excess water, are also typical. Usually protozoans have some form of locomotion; some use **pseudopodia,** some move by **cilia,** and some use **flagella** (Fig. 12.7). Sporozoans—such as *Plasmodium vivax,* which causes a common form of malaria—do not locomote at all.

### Observation: Protozoans

1. You may already have had the opportunity to observe *Euglena* in Laboratory 2. However, your instructor may want you to observe these organisms again.
2. If a video, CD-Rom, or film is available, watch it, and note the various forms of protozoans.
3. Prepare wet mounts or examine prepared slides of protozoans as directed by your instructor.
4. Note the means of locomotion for each protozoan in Figure 12.7.

### Figure 12.7  Protozoan diversity.

The protozoans are unicellular and heterotrophic by ingestion. Some protozoans move by pseudopods, cilia, or flagella. *a. Amoeba*; *b. Paramecium*; *c. Trypanosoma*.

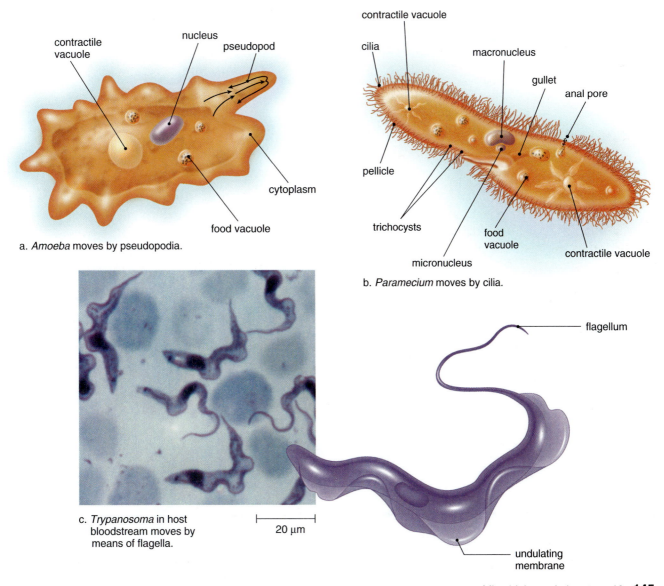

a. *Amoeba* moves by pseudopodia.

b. *Paramecium* moves by cilia.

c. *Trypanosoma* in host bloodstream moves by means of flagella.

20 μm

## Pond Water

Pond water typically contains various examples of protozoans and algae, the protists studied in this laboratory.

### Observation: Pond Water

1. Prepare a wet mount of a sample of pond water. Be sure to select some of the sediment on the bottom and a few strands of filamentous algae.
2. Identify the organisms you see by consulting Figure 12.8. *On the lines provided, write in the means of locomotion for each protist.* Those with chloroplasts are algae, and those without chloroplasts are protozoans. *Difflugia* and *Arcella* are amoebas. No line is provided for *Asplanchna, Philodina, Keratella,* and *Chaetonotus* because these are actually very small animals.

**Figure 12.8**  Microorganisms found in pond water.

| | | | | | |
|---|---|---|---|---|---|
| Amoeba | Euglena | Paramecium | Vorticella | Euplotes | Asplanchna |
| Tetrahymena | Stentor | Colpoda | Didinium | Philodina | Keratella |
| Chilomonas | Chlamydomonas | Difflugia | Arcella | Eudorina | Pandorina |
| Blepharisma | Chilodonella | Stylonychia | Dileptus | Chaetonotus | Oxytricha |

## Slime Molds

**Slime molds** are called the funguslike protists. They are saprotrophic like fungi, and they do form spores during some part of their life cycle.

There are two types of slime molds: cellular slime molds and plasmodial slime molds. **Cellular slime molds** usually exist as individual amoeboid cells, which aggregate on occasion to form a pseudoplasmodium. **Plasmodial slime molds** usually exist as a **plasmodium**, a fan-shaped, multi-nucleated mass of cytoplasm. The plasmodium of a plasmodial slime mold creeps along, phagocytizing decaying plant material in a forest or an agricultural field. During times unfavorable for growth, such as a drought, the plasmodium develops many sporangia. A **sporangium** is a reproductive structure that produces spores by meiosis. In some plasmodial slime molds, the spores become flagellated cells, and in others, they are amoeboid. In any case, they fuse to form a zygote that develops into a plasmodium (Fig. 12.9).

### Observation: Plasmodial Slime Mold

1. Obtain a plate of *Physarum* growing on agar. Examine the plate carefully under the dissecting microscope.

2. Describe what you see. _____

_____

**Figure 12.9** **Plasmodial slime molds.**
*a.* The plasmodium, multinucleated cytoplasm, creeps along forest floor. *b.* During sexual reproduction when conditions are unfavorable for growth, the diploid adult forms sporangia. Haploid spores germinate, releasing haploid amoeboid or flagellated cells that fuse to form a plasmodium.

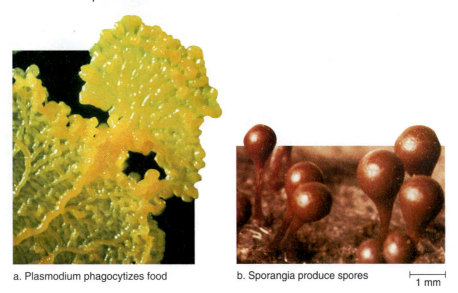

a. Plasmodium phagocytizes food        b. Sporangia produce spores        1 mm

## 12.3 Fungi

Fungi (kingdom Fungi) (Fig. 12.10) are saprotrophic in the same manner as bacteria. Both fungi and bacteria are often referred to as "organisms of decay" because they break down dead organic matter and release inorganic nutrients for plants. A fungal body, called a **mycelium,** is composed of many strands, called **hyphae** (Fig. 12.11). Sometimes, the nuclei within a hypha are separated by walls called septa.

Fungi produce windblown **spores** (small, haploid bodies with a protective covering) when they reproduce sexually or asexually.

**Figure 12.10  Diversity of fungi.**
*a.* Scarlet hood, an inedible mushroom. *b.* Spores exploding from a puffball. *c.* Common bread mold. *d.* A morel, an edible fungus.

c.

d.

**Figure 12.11  Body of a fungus.**
*a.* The body of a fungus is called a mycelium. *b.* A mycelium contains many individual chains of cells, and each chain is called a hypha.

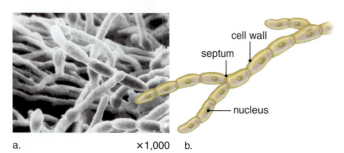

cell wall

septum

nucleus

a.          ×1,000    b.

## Black Bread Mold

In keeping with its name, black bread mold grows on bread and any other type of bakery goods. Notice in Figure 12.12 sporangia at the tips of aerial hyphae produce spores in both the asexual and sexual life cycles. A zygospore is diploid (2n); otherwise, all structures in the asexual and sexual life cycles of bread mold are haploid (n).

### Observation: Black Bread Mold

1. If available, examine bread that has become moldy. Do you recognize black bread mold on the bread?
2. Obtain a petri dish that contains living black bread mold. Observe with a dissecting microscope. Identify the mycelium and a sporangium.

**Figure 12.12** **Black bread mold.**
The mycelium of this mold, which uses sporangia to produce windblown spores, lives on bread.

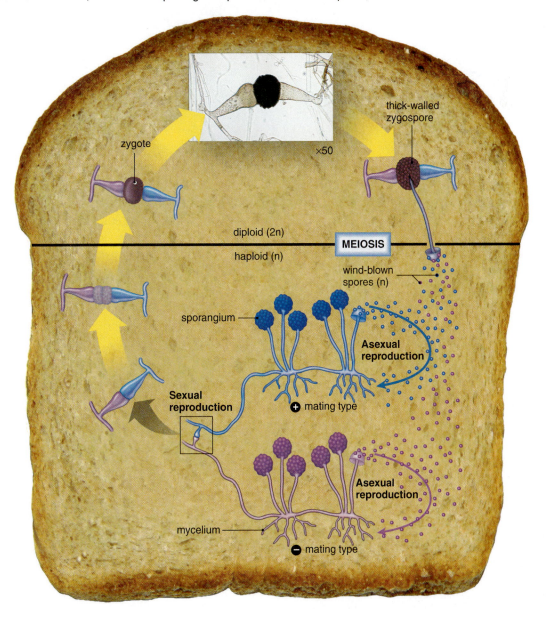

**3.** View a prepared slide of *Rhizopus,* using both a dissecting microscope and the low-power setting of a light microscope. The absence of cross walls in the hyphae is an identifying feature of zygospore fungi. *Label the mycelium and zygospore in Figure 12.13b.*

**Figure 12.13** **Microscope slides of black bread mold.**
*a.* Asexual life cycle. *b.* Sexual life cycle.

a.

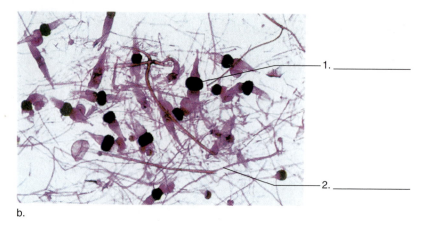

1. _____

2. _____

b.

## Club Fungi

**Club fungi** are just as familiar as black bread mold to most laypeople because they include the mushrooms. A gill mushroom consists of a stalk and a terminal cap with gills on the underside (Fig. 12.14). The cap, called a basidiocarp, is a fruiting body that arises following the union of + and – hyphae. The gills bear basidia, club-shaped structures where nuclei fuse, and meiosis occurs during spore production. The spores are called basidiospores.

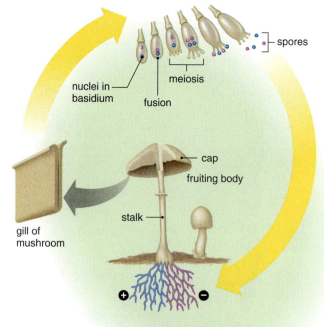

spores

nuclei in basidium

meiosis

fusion

cap

fruiting body

stalk

gill of mushroom

**Figure 12.14** **Sexual reproduction produces mushrooms.**
Fusion of + and – hyphae tips results in hyphae that form the mushroom (a fruiting body). The nuclei fuse in clublike structures attached to the gills of a mushroom, and meiosis produces spores.

**150** Laboratory 12  Microbiology

12–14

## Observation: Mushrooms

1. Obtain an edible mushroom—for example, *Agaricus*—and identify as many of the following structures as possible:

   a. **Stalk:** The upright portion that supports the cap
   b. **Annulus:** A membrane surrounding the stalk where the immature (button-shaped) mushroom was attached
   c. **Cap:** The umbrella-shaped basidiocarp of the mushroom
   d. **Gills:** On the underside of the cap, radiating lamellae on which the basidia are located
   e. **Basidia:** On the gills, club-shaped structures where basidiospores are produced
   f. **Basidiospores:** Spores produced by basidia

2. View a prepared slide of a cross section of *Coprinus*. Using all three microscope objectives, look for the gills, basidia, and basidiospores.

3. Can you see individual hyphae in the gills? _____

4. Are the basidiospores inside or outside of the basidia? _____

5. What type of nuclear division does the zygote undergo to produce the basidiospores? _____
   _____

6. Can you suggest a reason for some of the basidia having fewer than four basidiospores? _____
   _____

7. What happens to the basidiospores after they are released? _____
   _____

## Fungi and Human Diseases

Fungi cause a number of human diseases. Oral thrush is a yeast infection of the mouth common in newborns and AIDS patients (Fig. 12.15a). Ringworm is a group of related diseases caused by the fungi *tinea.* The fungal colony grows outward, forming a ring of inflammation (Fig. 12.15b). Athlete's foot is a form of tinea that affects the foot, mainly causing itching and peeling of the skin between the toes (Fig. 12.15c).

**Figure 12.15  Human fungal diseases.**
*a.* Thrush, or oral candidiasis, is characterized by the formation of white patches on the tongue. *b.* Ringworm and *c.* athlete's foot are caused by *Tinea* spp.

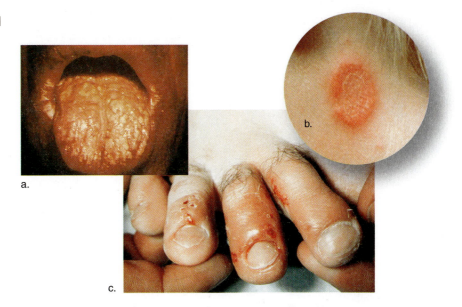

_____ 1. What role do bacteria and fungi play in ecosystems?

_____ 2. What type of semisolid medium is used to grow bacteria?

_____ 3. What is the scientific name for spherical bacteria?

_____ 4. It is sometimes said that diatoms live in what kind of "houses"?

_____ 5. What type of nutrition do algae have?

_____ 6. Name a colonial alga studied today.

_____ 7. Gram-positive bacteria have a thick layer of what substance in their cell walls?

_____ 8. What color are Gram-negative bacteria following Gram staining?

_____ 9. Once called the blue-green algae, cyanobacteria are now classified as what?

_____ 10. What do you call the projection that allows amoeboids to move and feed?

_____ 11. What do you call the multinucleate stage of the plasmodial slime mold?

_____ 12. The stalk and cap of a mushroom that rise above the substratum are termed what?

_____ 13. What type of nutrition do fungi have?

_____ 14. What do fungi produce during both sexual and asexual reproduction?

## Thought Questions

15. Why aren't all the organisms studied today in the same domain?

16. Fungi always reproduce by producing haploid spores during both asexual and sexual reproduction. In general, how does sexual reproduction differ from asexual reproduction among fungi?

# 13

# Seedless Plants and Seed Plants

## Introduction

Your study of plant evolution in this laboratory will emphasize four groups of plants: the mosses; the ferns; the gymnosperms, represented by the pine tree; and the angiosperms, the flowering plants. While each of these groups of plants are successfully adapted to living on land, adaptation to the land environment is best demonstrated by the flowering plants (Fig. 13.1). The number and kinds of flowering plants is much greater than that of all the other groups of plants.

Adaptation to a land environment includes the ability to prevent excessive loss of water into the atmosphere; to obtain and transport water and nutrients to all parts of the plant; to support a large body against the pull of gravity; and to reproduce without dependence on external water.

With regard to human beings, consider that skin protects us from drying out, blood transports water and nutrients about the body, the skeleton supports us, males have a penis for delivering flagellated sperm to the female, and the embryo and fetus are protected from drying out within the uterus of the female.

**Figure 13.1  The flowering plants.**
The flowering plants provide us with much of our food including grains, potatoes, and beans. Here, we see the flower and the fruit of a watermelon plant.

## Figure 13.2  Evolution of plants.

The evolution of plants is marked by four significant events: protection of the embryo, evolution of vascular tissue, evolution of the seed, and evolution of the flower.

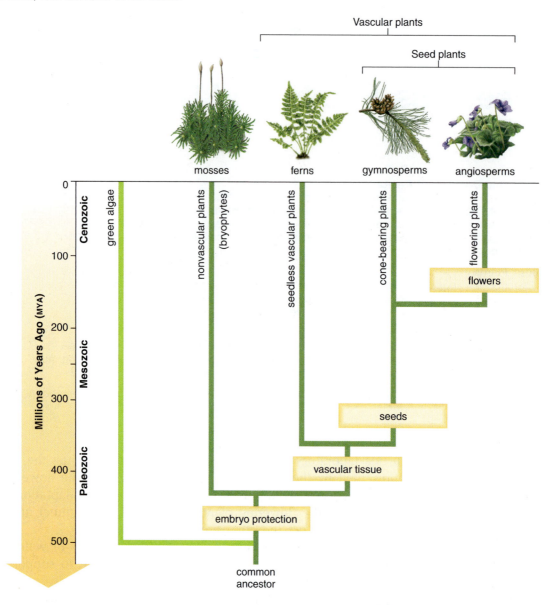

## Figure 13.3  Representatives of the four major groups of plants.

a. In mosses, the embryo is protected by a special structure.

b. A fern has vascular tissue.

c. In a conifer, seeds disperse offspring.

d. In flowering plants, the seeds are enclosed in fruits.

# 13.1 Evolution and Diversity of Plants

In Figure 13.2, note the four events that mark the evolution of plants. Protection of the embryo is first seen in mosses (Fig. 13.3). The next event, evolution of vascular tissue, is first seen in ferns. Vascular tissue not only allows transport of water and nutrients, it also provides an internal skeleton that opposes the force of gravity. Conifers are the first seed plants. The last event, evolution of flowers, allows the seeds to be protected within a fruit. Fruits aid dispersal of seeds.

## Alternation of Generations

All plants have a life cycle known as alternation of generations (Fig. 13.4). In this life cycle, there are two mature stages, known as the sporophyte and the gametophyte. The **sporophyte,** the 2n generation, produces spores, by the process of meiosis, in structures called sporangia (sing., sporangium). A **spore** is a haploid reproductive cell that produces a new generation that is also haploid. Spores develop into the gametophyte. The **gametophyte,** the n generation, produces gametes that later fuse to form a zygote. A zygote develops into the sporophyte, completing the cycle.

In plants, one generation is **dominant,** meaning that it lasts longer and is the generation we refer to as the plant. The gametophyte is dominant in mosses, and the sporophyte is dominant in ferns, gymnosperms, and angiosperms. In plants with dominant sporophytes, the sporophyte has vascular (transport) tissue that conducts water from the roots to the leaves. Is it beneficial for a

sporophyte, the generation that has vascular tissue, to be dominant? _____

Why? _____

## Figure 13.4  Alternation of generations.

In the alternation of generations life cycle of plants, the sporophyte is the 2n generation that produces spores by meiosis. The gametophyte is the n generation that produces the gametes. When the gametes fuse, a new sporophyte comes into being.

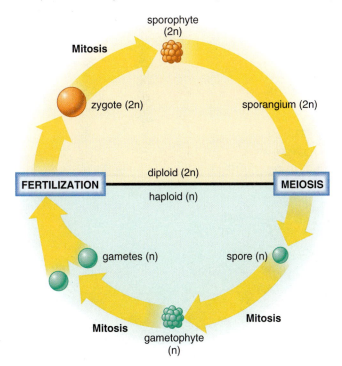

# 13.2 Seedless Plants

Mosses and ferns are both seedless plants. Mosses and their relatives, called the bryophytes, are low-lying plants, called the nonvascular plants because they lack vascular tissue. Ferns, characterized by large leaves, do have vascular tissue, but even so, share other characteristics with the bryophytes. For example, they are both seedless plants.

## Mosses

The bryophytes (mosses and liverworts) were the first plants to live on land. The gametophyte is dominant in these nonvascular plants. The gametophyte produces eggs within archegonia and swimming sperm in antheridia (Fig. 13.5). The bryophytes are dependent on external moisture to ensure fertilization because the sperm must swim to the egg. However, the zygote developing within the archegonia is protected from drying out. The bryophytes have another adaptation to life on land in that the spores, produced by the dependent sporophyte, are windblown. The spores disperse the gametophyte.

**Figure 13.5** **Life cycle of the moss.**
In mosses, the gametophyte consists of leafy shoots that produce flagellated sperm within antheridia and eggs within archegonia. Following fertilization, the sporophyte, consisting of a stalk and capsule (a sporangium), is dependent on the gametophyte.

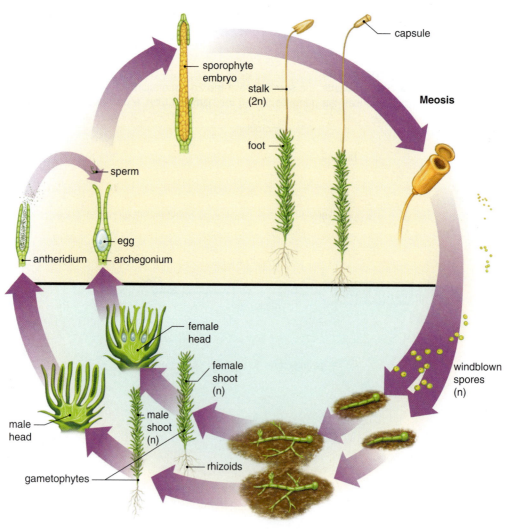

## Life Cycle of Mosses

Study the life cycle of the moss (see Fig. 13.5) and find the gametophyte. Examine a living gametophyte or a plastomount of this generation. Describe its appearance. _____ _____ Considering that this is the

generation we refer to as the "moss," what generation is dominant in mosses? _____

Describe the sporophyte. _____

### Observation: Moss Life Cycle

1. Study the prepared slide of the female shoot head (top of female shoot) and the male shoot head (top of the male shoot). Find an **archegonium** in a female shoot head and locate an egg in at least one of these. Find an **antheridium** in a male shoot head. An antheridium is filled with many flagellated sperm. When sperm produced by the antheridia swim in a film of water to the eggs in the archegonia, zygotes result. A zygote develops into a new sporophyte.

2. Examine the plastomount of a shoot with a sporophyte attached. The sporophyte is dependent on the female shoot. Why female? _____

3. Examine a slide of a longitudinal section through the sporophyte of the moss. Identify the stalk and sporangium. What is being produced in the sporangium? _____ By what process? _____

### The Life Cycle of a Moss

1. Which generation is haploid? _____ Which is diploid? _____
   Which generation is dominant in mosses? _____ Which generation is dependent? _____

2. Is there any evidence of vascular tissue in the moss sporophyte? _____

3. When spores germinate, what generation begins to develop? _____

4. Why is it proper to say that, in the moss, spores disperse the plant? _____ _____

5. By what means are spores disseminated? _____

### Adaptation of Mosses to the Land Environment

Which of these is an indication that mosses are well adapted to life on land? Write "yes" if the feature is an adaptation to land and "no" if the feature is not an adaptation to living on land.

Lack of vascular tissue. _____

Body covered by a cuticle, a protection against drying out. _____

Flagellated sperm. _____

Egg and embryo protected by female shoot. _____

Spores are windblown. _____

## Ferns

All the other plants to be studied in this laboratory are vascular plants. Ferns are seedless vascular plants in which the dominant sporophyte possesses vascular tissue. The windblown spores develop into an independent gametophyte that is water dependent because it lacks vascular tissue and also because it produces flagellated sperm. The sperm must swim from the antheridia to the archegonia, where the eggs are produced. The zygote, protected from drying out within an archegonium, develops directly into the sporophyte.

### *Observation: Fern Life Cycle*

1. Study the life cycle of the fern (Fig. 13.6) and find the sporophyte and gametophyte. This large, complexly divided sporophyte leaf is known as a **frond.** Fronds arise from an underground stem called a **rhizome.** Examine a preserved specimen of a frond and on the underside, notice brownish **sori,** each one a cluster of many sporangia (Fig. 13.7). What is being produced in the sporangia? _____ Considering that it is this generation that we call the fern, what generation is dominant in ferns? _____

**Figure 13.6  Fern life cycle.**
The frond is the dominant sporophyte generation that produces windblown spores. A spore gives rise to an independent gametophyte, called a prothallus. The prothallus produces gametes. When the gametes fuse, a new sporophyte begins to develop.

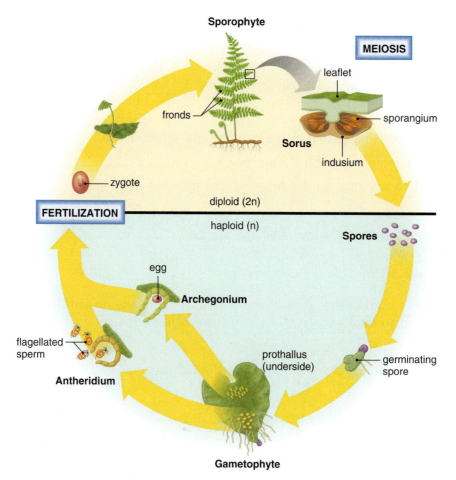

**Figure 13.7**  Underside of frond leaflets showing sori.

sorus

2.  Examine a prepared slide of young sporangium, cross section. Study the slide carefully and locate the fern leaf and a sorus (pl., sori). Within a sorus, find the sporangia and spores. Notice the **indusium,** a shelflike structure that protects the sporangia until they are mature.
3.  Examine a plastomount showing the fern life cycle. Notice a portion of the frond with sori and a small heart-shaped structure, the **prothallus.** The latter is the gametophyte generation of the fern. Most persons do not realize that this structure exists as a part of the fern life cycle. What is the function of this structure? _____
4.  Examine a whole mount slide of a fern prothallium-archegonia. What is being produced inside an archegonium? _____ If you focus up and down very carefully on an archegonium, you may be able to see an egg inside.
5.  Examine a whole mount slide of a fern prothallium-antheridia. What is being produced inside the antheridia? _____ When sperm produced by the antheridia swim to the archegonia in a film of water, what results? _____ The latter develop into what generation? _____

## The Life Cycle of a Fern

1.  Is either generation in the fern dependent for any length of time on the other generation? _____
2.  Which generation is dispersed in ferns? _____ How? _____

## Adaptation of Ferns to a Land Environment

1.  List one additional way in which the fern is adapted to life on land, when you compare it to the moss. _____
2.  List one characteristic of the fern illustrating that sexual reproduction is not adapted to a land environment. _____

## 13.3 Seed Plants

Seed plants (gymnosperms and angiosperms) are further adapted to live and reproduce on land. The dominant sporophyte contains vascular tissue, which not only transports water but also serves as an internal skeleton, allowing these plants to oppose the force of gravity. For example, all of today's trees are seed plants.

Seed plants are no longer dependent on external water to ensure fertilization. To understand the mechanism by which this has come about, it is necessary to know that seed plants have two types of sporangia, termed **microsporangia** and **megasporangia** (Fig. 13.8). In microsporangia, microspore mother cells produce microspores by meiosis and each develops into a male gametophyte generation, the **pollen grain.** During **pollination,** pollen grains are dispersed to the vicinity of female gametophytes.

In a megasporangium inside an **ovule,** a megaspore mother cell produces megaspores by meiosis. Only one of these develops into a female gametophyte that produces an egg. Following fertilization, the 2n zygote is still within the ovule, which becomes a seed. The seed is dispersed to a new location.

### Seed Plants

1. Seed plants have two dispersal events. They are dispersal of _____ and dispersal of _____. In gymnosperms, pollen grains are dispersed by the wind; and in angiosperms, dispersal is sometimes by the wind and sometimes by insects or other animals.

2. During dispersal of seeds, what generation is dispersed to a new location? _____

### Figure 13.8  Dispersal in seed plants.

(*top*) Pollen grains are produced in microsporangia and are dispersed during pollination. After reaching the female gametophyte, a pollen grain germinates and has a pollen tube. A sperm travels down the pollen tube to the egg. (*bottom*) An ovule, at first, contains a megasporangium; then a megaspore, which develops into a female gametophyte that produces an egg. Following fertilization, the ovule becomes the seed, which is dispersed to a new location.

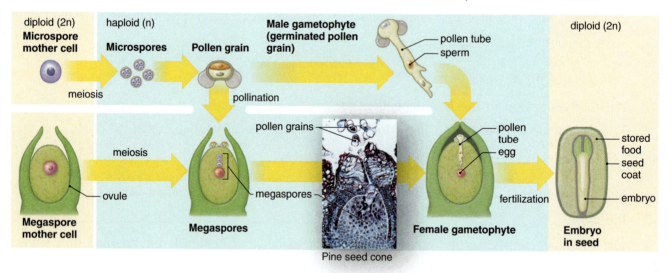

## Gymnosperms

The gymnosperms are usually evergreen trees in which the sporangia are found on **cones** (Fig. 13.9 and Fig. 13.10). The **conifers** are by far the largest group of gymnosperms. Pines, spruces, firs, cedars, hemlocks, redwoods, and cypresses are all conifers. The pine has been chosen as representative of these plants.

**Figure 13.9** Life cycle of pine.

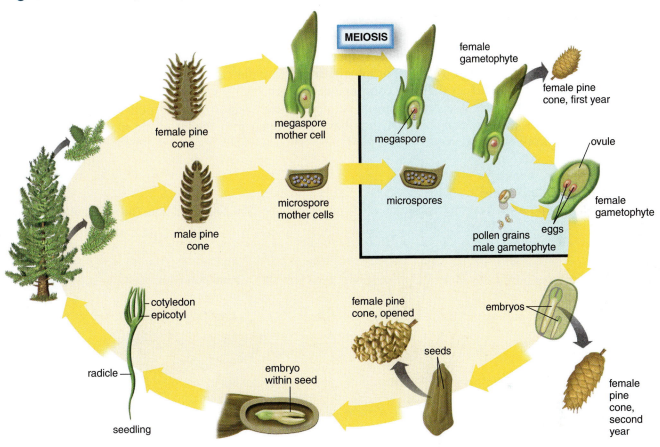

**Figure 13.10** Conifers.

*a.* Vast areas of northern temperate regions are covered in evergreen coniferous forests. The tough, needlelike leaves of pines conserve water because they have a thick cuticle and recessed stomata. *b.* Conifers bear two types of cones: pollen (male) cones and seed (female) cones. *c.* The seed cones of junipers are fleshy.

a. A northern coniferous forest

b. Cones of lodgepole pine, *Pinus contorta*

c. Juniper, *Juniperus*, seed cones are fleshy

## Life Cycle of Pine Trees

1. Describe the sporophyte in the life cycle of the pine tree. _____
2. The sporophyte produces pollen (male) cones and seed (female) cones. Two microsporangia are located on the underside of each scale from the pollen cone. Here, microspore mother cells produce microspores that develop into pollen grains, the male gametophyte generation. Two ovules (megasporangia) are located on the underside of each scale from the seed cone. Here, megaspore mother cells produce four cells, one of which survives and becomes a megaspore.

### Observation: Pine Cones

1. Observe the pollen and seed cones on display (Fig. 13.11). Compare their relative size and structure. Remove a single scale (sporophyll) from the male cone and from the seed cone, which has been heated so that the cone opens. Observe with a dissecting microscope. Note the two microsporangia located on the lower surface of each scale from the pollen cone. Note also the two ovules (megasporangia) that may have developed into seeds on the upper surface of each scale from the seed cone.

**Figure 13.11**  **Pine cones.**
*a.* The scales of pollen cones bear microsporangia, where microspores become pollen grains. *b.* The scales of seed cones bear ovules that develop into winged seeds.

a. Pollen cones

microsporangium

scale

b. Seed cone

seed

wing

scale

2. Examine the prepared slide of a longitudinal section through a mature pollen cone and label Figure 13.12. On the lower surface of each scale are the microsporangia in which the microspore mother cells produce microspores. Each microspore develops into a male gametophyte, a pollen grain. Under high power, focus on a pollen grain and note the external wings and interior cells. One of these will divide to produce two cells, one of which is the sperm nucleus and the other of which forms the pollen tube, through which a sperm travels to the egg.

**Figure 13.12** Pine pollen cone.
Pollen cones bear *a.* microsporangia in which microspores develop into pollen grains. *b.* Enlargement of pollen grains.

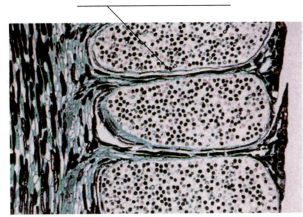

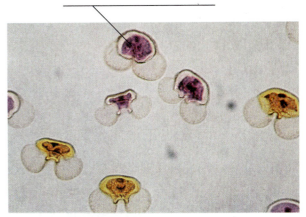

a. Longitudinal section through pine pollen cone, showing pollen grains within microsporangia.

b. Enlargement of pollen grains.

3. Examine the prepared slide of a longitudinal section through a seed cone and label Figure 13.13. In some ovules, you will see a megaspore mother cell surrounded by nutritive cells. You may also be able to find pollen grains just outside or within the integuments of an ovule. At about the time the megaspore mother cell is undergoing meiosis, the scales swell and open to allow the wind-dispersed pollen grains to enter. The megaspore undergoes a series of mitotic divisions and develops into the female gametophyte that contains two archegonia, each of

which encloses a single large egg. What generation is now within the ovule? _____

**Figure 13.13** Seed cone.
Seed cones bear ovules, each of which at one time contains a megasporangium shown here in longitudinal section. Note pollen grains near the entrance.

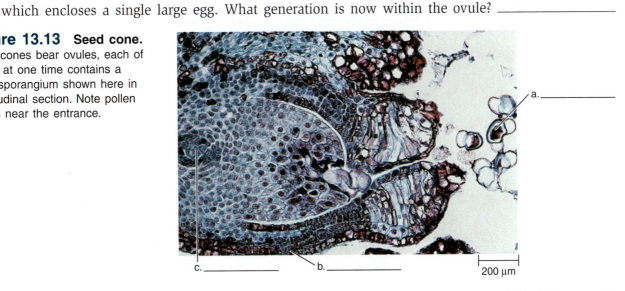

a. _____

c. _____  b. _____  200 µm

4. Following fertilization, a seed contains the embryonic plant and nutrient material within a seed coat. If available, examine pine seeds. The seeds of gymnosperms are windblown. In seed plants, seeds disperse the sporophyte. The word gymnosperm means "naked seeds," and the seeds themselves are not covered. If you wish, dissect a seed and, with the help of a hand lens, attempt to find the embryo.

## Angiosperms

The angiosperms are the flowering plants. The flower contains the microsporangia and megasporangia. The word angiosperm means covered seeds, and these plants have seeds enclosed in fruits. The success of angiosperms is exemplified by their large number compared to that of the other plants.

### Life Cycle of Flowering Plants

Study the diagram of a flowering plant life cycle (Fig. 13.14) and find the sporophyte generation. Describe the sporophyte, if the flowering plant is a deciduous tree (loses leaves in fall).

Describe the sporophyte of a garden plant. _____

**Figure 13.14** **Life cycle of flowering plants.**

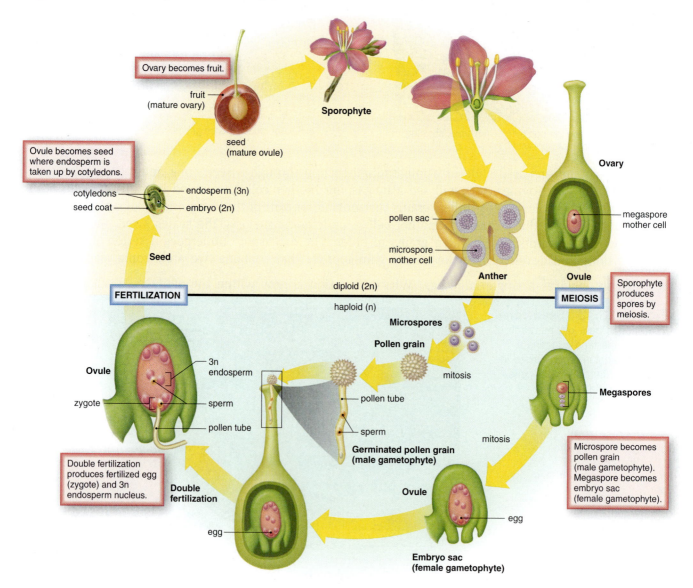

Figure 13.14 shows that each mother cell within the anther undergoes meiotic cell division to produce four haploid cells, the microspores. Each microspore divides mitotically to give a male gametophyte, a pollen grain. Following **pollination,** the transfer of pollen from the anther to the stigma, a pollen grain germinates and produces a long **pollen tube** in which two sperm travel to reach the ovule.

Inside each ovule, a megaspore mother cell undergoes meiosis to produce four megaspores, three of which disintegrate. The one remaining undergoes mitosis to give a multicelled female gametophyte. Flowering plants practice **double fertilization** because the other sperm joins two polar nuclei to form endosperm (3n), which serves as food for the developing embryo. Now the ovule becomes a seed. A seed contains the embryonic sporophyte and stored food within a seed coat. The seeds of angiosperms are covered by the ovary. The ovary becomes the fruit. All angiosperms produce fruit.

## Observation: Flower

1. Identify these parts of a flower on a model using Figure 13.15.

   **Sepal:** the outermost set of modified leaves, collectively termed the **calyx.** The sepals are green in most flowers.

   **Petal:** the inner leaves that collectively comprise the **corolla.** The petals tend to be brightly colored.

   **Stamen:** comprised of a swollen terminal **anther** and a slender supporting filament.

   **Carpel:** comprised of a swollen basal **ovary;** a long, slender **style;** and a terminal **stigma.**

   **Ovule:** structure within ovary that contains the megasporangium and develops into the seed.

**Figure 13.15  Anatomy of typical flower.**

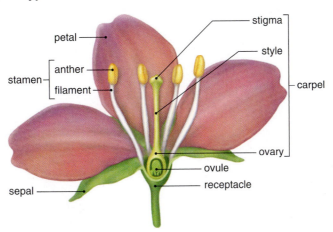

2. Carefully inspect a living flower. Remove the sepals and petals by breaking them off at the base. Are the stamens taller than the carpel? _____ Would self-fertilization be possible in this flower? _____

3. Remove a stamen and touch the anther to a drop of water on a slide. If nothing comes off in the water, crush the anther a little to squeeze out some of its contents. Place a coverslip on the drop and observe with low and high powers of the microscope. The somewhat spherical cells with thick walls are pollen grains. How is pollen dispersed? _____ Flowering plants provide nectar to insects, whose mouth parts are adapted to acquiring the nectar from this particular species of plant. This is a mutualistic relationship. What does a pollinator do for the plant? _____

4. Remove the carpel by cutting it free just below the base. Make a series of thin cross sections through the ovary. The ovary is hollow and in this cavity there are small, nearly spherical bodies much larger than pollen grains. These are ovules. Remove an ovule that may be still attached by a stalk to the ovary wall. Place the ovule on a drop of water on a slide; cover with a cover glass; press firmly on the top of the cover glass with a clean eraser so as to smear the ovule into a thin mass. Observe with the microscope. Do you see cells within the ovule? _____

5. At maturity, an angiosperm seed contains the embryo and possibly some nutrient material in the form of endosperm covered by a seed coat. A fruit is a ripened ovary, together with any accessory flower parts that may be associated with the ovary. If available, examine a pea pod and an apple. Find the remnants of the flower parts still attached to the fruit. Open the pea pod and slice the apple. Which portion of each should be associated with the ovules; which with the ovary? Label flower remnants, fruit, and seed in the following diagram. Can you think of a biological advantage to producing fruits? _____ To producing a fleshy fruit?

_____

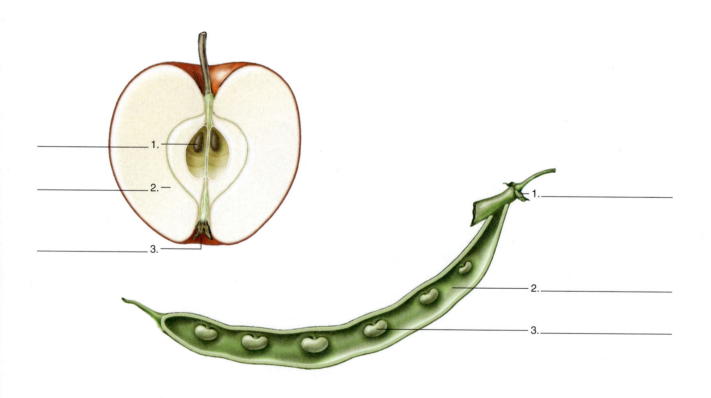

## Compare the Life Cycle of a Pine Tree to that of a Flowering Plant

Complete the following table:

| | Dominant Generation | Vascular Tissue (Present or Absent) | Dispersal of Sporophyte | Fruit (Present or Absent) |
|---|---|---|---|---|
| Conifer | | | | |
| Flowering plant | | | | |

## Adaptation of Seed Plants to a Land Environment

1. Do seed plants have vascular tissue that transports water and nutrients to all parts of the plant? _____

2. Can seed plants support a large body against the pull of gravity? _____

3. Do seed plants reproduce without dependence on external water? _____

## Comparison of Moss, Fern, Pine, and Flowering Plants

1. The preceding diagram tells you that the size of the _____ became progressively _____ as plants became adapted to live on land. Why is it suitable for this generation to be dependent on the sporophyte? _____

2. Which plants disperse spores? _____

3. Which plants have male and female gametophytes and disperse seeds? _____

4. Why is it appropriate to refer to "bees and plants" to explain sexual reproduction? _____

_____

_____  1. The sporophyte is the haploid or diploid generation?

_____  2. What generation is in a seed?

_____  3. Which generation is the dependent one in mosses?

_____  4. What structure is used for dispersal in seedless plants?

_____  5. Which groups of plants have flagellated sperm?

_____  6. Do bryophytes or ferns have vascular tissue?

_____  7. What generation is dominant in ferns, gymnosperms, and angiosperms?

_____  8. Which group of plants studied has vascular tissue but uses spores for dispersal?

_____  9. What kind of spore becomes a pollen grain in seed plants?

_____  10. The carpel should be associated with the development of which gametophyte, male or female?

_____  11. The ovule becomes the seed, while the ovary becomes what?

_____  12. The gametophyte in seed plants is dependent on or independent of the sporophyte?

## Thought Questions

13. A pine tree, unlike a fern, is able to reproduce sexually in a dry environment. Explain.

14. What is the difference between pollination and fertilization?

# 14

# Plant Anatomy and Growth

## Learning Objectives

### 14.1 Plant Organs
- Distinguish between the root system and the shoot system of a plant.
- Identify the external anatomical features of a flowering plant.
- State the functions of a leaf, a stem, and a root.
- List five differences between monocots and eudicots.

### 14.2 Organization of Roots
- Name the zones of a eudicot root tip.
- Identify a cross section and the specific tissues of a eudicot root.

### 14.3 Xylem Transport
- Explain the continuous water column in xylem.

### 14.4 Organization of Stems
- Identify a cross section and the specific tissues of herbaceous eudicot and herbaceous monocot stems and a woody stem.
- Distinguish between primary and secondary growth of a stem.
- Explain the occurrence of annual rings, and determine the age of a tree from a trunk cross section.

### 14.5 Organization of Leaves
- Distinguish between a monocot and a eudicot leaf.
- Identify a cross section and the specific tissues of a eudicot leaf.

## Introduction

Despite their great diversity in size and shape, flowering plants all have three vegetative organs that have nothing to do with reproduction: the leaf, the stem, and the root. Leaves carry on photosynthesis and thereby, produce the nutrients that sustain a plant. A stem usually supports the leaves so that they are exposed to sunlight. Roots anchor a plant and absorb water and minerals from the soil.

Each of these organs contains various tissues, arranged differently depending on whether a flowering plant is a monocot or a eudicot. The arrangement of tissues is distinctive enough that you should be able to identify the plant as a monocot or eudicot when examining a slide of a leaf, stem, or root.

Another way of grouping plants is according to whether they are herbaceous or woody. All flowering trees are woody; their stems contain wood. Many flowering garden plants and all grasses are herbaceous (nonwoody). Herbaceous plants have only **primary growth,** which increases their height. Woody plants have both primary and secondary growth. **Secondary growth** increases the girth of a tree. Only eudicots are ever woody.

Xylem is the vascular tissue that transports water from the roots to the leaves. The cohesion-tension model of xylem transport explains how the continuous column of water in xylem is able to rise to the top of a tall tree. During transpiration, water evaporates through openings in leaves called stomata (sing., stoma). This creates a tension that pulls the water column up because of the cohesive property of water.

# 14.1 Plant Organs

Figure 14.1 shows that a plant has a root system and a shoot system. The **shoot system** consists of the stem and leaves. The **root system** consists of a primary root and all of its lateral (side) roots.

## Observation: A Living Plant

### Shoot System

What is the primary function of the shoot system? _____

_____

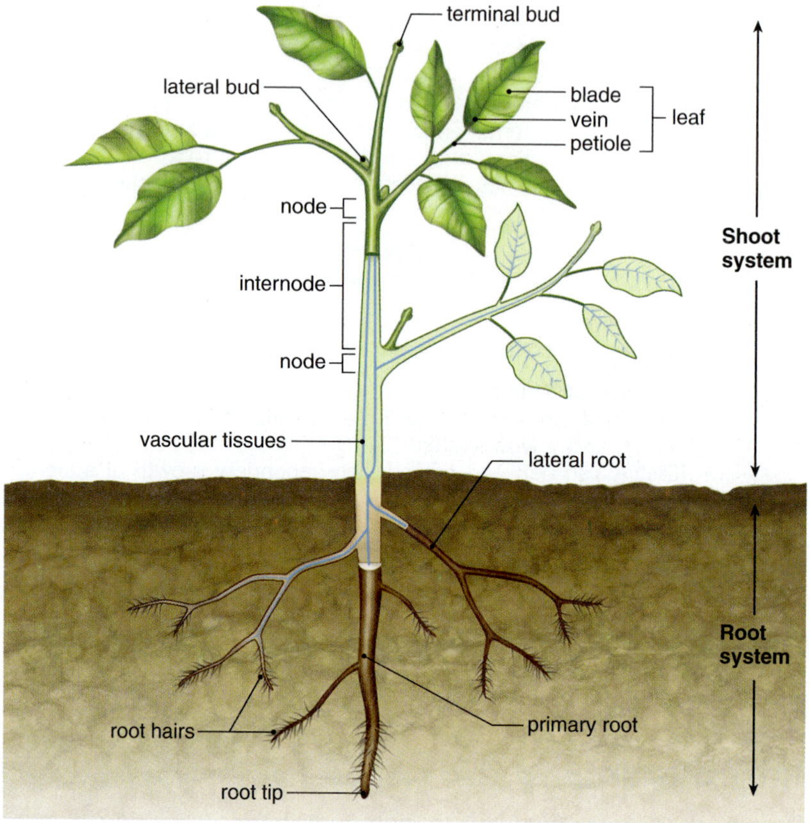

### The Leaves

1. Describe the **blade.**

   _____

2. Describe the **petiole.**

   _____

### The Stem

1. Observe the **stem.** Locate a **node** and an **internode.**
2. Measure the length of the internode in the middle of the stem. Does the internode get larger or smaller toward the apex of the stem?

   _____

   Toward the roots? _____
   Based on the fact that a stem elongates as it grows, explain
   your observation. _____

   _____

**Figure 14.1  The body of a plant.**
A plant has a root system, which extends below ground, and a shoot system, composed of the stem and leaves.

terminal bud
lateral bud
blade
vein
petiole
leaf
node
internode
node
vascular tissues
lateral root
root hairs
primary root
root tip
Shoot system
Root system

3. Where is the **terminal bud** of a stem? _____ Where is the **lateral**

   bud? _____ Buds produce new growth because they contain a plant tissue called meristem. **Meristem** is a plant tisue consisting of undifferentiated cells that can ever divide to produce cells that later become specialized.

### Root System

Observe the root system of a living plant if the root system is exposed. Does the plant have a taproot system—that is, a main root many times larger than the lateral roots? _____ Or does the plant have a fibrous root system—that is, all the roots approximately the same size? _____ What is the primary function of the root system? _____

## Figure 14.2 Flowering plants are either monocots or eudicots.

Five features are typically used to distinguish monocots from eudicots: number of cotyledons; arrangement of vascular tissue in roots, stems, and leaves; and number of flower parts.

**Monocots**

Seed: one cotyledon in seed

Root: root xylem and phloem in a ring

Stem: vascular bundles scattered in stem

Leaf: leaf veins form a parallel pattern

Flower: flower parts in threes and multiples of three

**Eudicots**

Seed: two cotyledons in seed

Root: root phloem between arms of xylem

Stem: vascular bundles in a distinct ring

Leaf: leaf veins form a net pattern

Flower: flower parts in fours or fives and their multiples

## Monocots Versus Eudicots

Flowering plants are classified into two groups: **monocots** and **eudicots.** In this laboratory, you will be studying the differences between monocots and eudicots with regard to the leaves, stems, and roots, as noted in Figure 14.2.

### Experimental Procedure: Monocot Versus Eudicot

1. Observe the leaves of the plant you are studying. Based on Figure 14.2, is this plant a monocot or a eudicot? _____ Explain. _____
2. Observe any other available types of leaves, and note in Table 14.1 the name of the plant and whether it is a monocot or a eudicot.

| Table 14.1 | Monocots Versus Eudicots | |
|---|---|---|
| Name of Plant | Organization of Leaf Veins | Monocot or Eudicot? |
| 1 | | |
| 2 | | |
| 3 | | |
| 4 | | |

# 14.2 Organization of Roots

First you will study a eudicot root tip in longitudinal section and then a eudicot root in cross section. You will also examine the root hairs of a living plant.

## Observation: Eudicot Root Tip

1. Obtain a model and/or a slide of a eudicot root tip.
2. With the help of Figure 14.3, identify

   **Root cap:** covers the growing root tip, which contains root apical meristem. What is the function of the root cap? _____

   _____

   **Zone of cell division:** where new cells are being produced.

   **Zone of elongation:** in this zone, rows of newly produced cells elongate. Which two zones are responsible for growth of the root tip? _____

   _____

   **Zone of maturation:** where the cells are specialized to carry on a particular function. You can recognize the zone of maturation because of the presence of root hairs. **Root hairs** increase the area for absorption of what by a root?

   _____

**Figure 14.3  Eudicot root tip.**
In longitudinal section, the root cap is followed by the zone of cell division, zone of elongation, and zone of maturation.

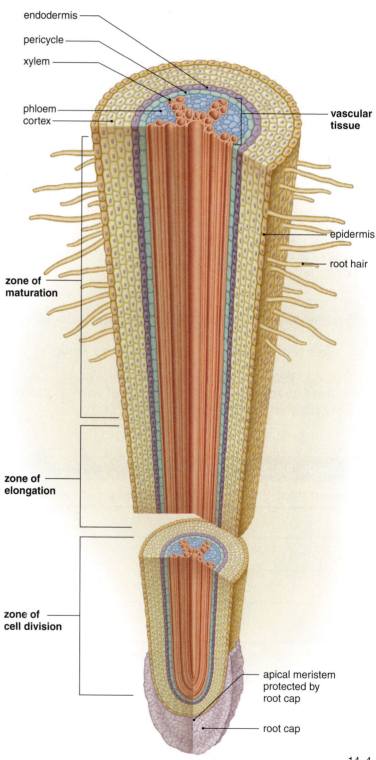

endodermis

pericycle

xylem

phloem

cortex

vascular tissue

epidermis

root hair

zone of maturation

zone of elongation

zone of cell division

apical meristem protected by root cap

root cap

## Figure 14.4 Eudicot root cross section.

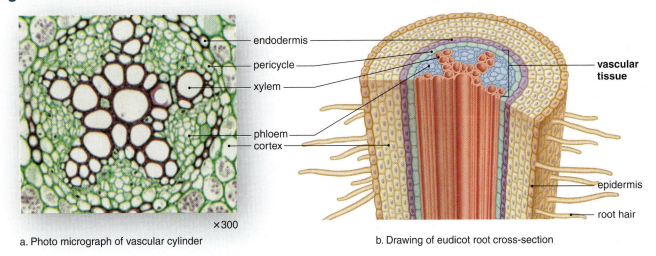

a. Photo micrograph of vascular cylinder ×300

endodermis
pericycle
xylem
phloem
cortex

vascular tissue
epidermis
root hair

b. Drawing of eudicot root cross-section

## Observation: Eudicot Root Slide

1. Obtain a slide of a eudicot root such as *Ranunculus* (buttercup) root, in cross section.
2. Identify

    **Epidermis:** the outermost layer of small cells that gives rise to **root hairs.** Notice in Figure 14.4 that the epidermis is a single layer of cells.

    **Cortex:** just inside the epidermis. How many layers of thin-walled cells are present? _____ The cortex stores the products of photosynthesis. Label starch grains in Figure 14.4*a*.

    **Endodermis:** a single layer of cells following the cortex. Endodermal cells have a waxy layer, called the Casparian strip, on four sides. This means that water and minerals have to pass through endodermal cells and not around them. For this reason, the endodermis regulates what materials enter a plant through the root.

    **Pericycle:** a layer one or two cells thick, just inside the endodermis. Lateral roots originate from this tissue.

    **Vascular tissue** (xylem and phloem). **Xylem** has several "arms" that extend like the spokes of a wheel. This tissue conducts water and minerals from the roots to the stem. **Phloem** is located between the arms of xylem. Phloem conducts organic nutrients from the leaves to the roots and other parts of the plant.

3. With the help of Figure 14.4, trace the path of water as it crosses a root from a root hair to

    xylem: _____

    _____

## Observation: Root Hairs

1. Obtain a young germinated seedling, and float it in some water in a petri dish while you observe it.
2. Use the binocular dissecting microscope, and locate the root tip. Note the root cap and the region where the root hairs have formed on the root surface. What proportion of the root has

    root hairs? _____
3. Remove the root from the seedling, and make a wet mount of the root, using 0.1% neutral red.
4. Observe your slide under the microscope. Does every epidermal cell have a root hair? _____

    How do root hairs aid absorption? _____

## 14.3 Xylem Transport

Xylem (Fig. 14.5), which transports water from the roots to the leaves, contains two types of conducting cells: tracheids and vessel elements. Both types of conducting cells are hollow and nonliving, the vessel elements are larger, they lack transverse end walls, and they are arranged to form a continuous pipeline for water and mineral transport. The water column in xylem is continuous because water molecules are cohesive (they cling together) and because water molecules adhere to the sides of xylem cells. Therefore, water evaporation from leaf surfaces creates a negative pressure that pulls water upward.

### Experimental Procedure: Xylem Transport

1. Using an eye dropper, place a small amount of *red-colored water* in two beakers. Label one beaker "wet" and the other beaker "dry."
2. Transfer a stalk of *celery* (which was cut and then immediately placed in a container of water) into the "wet" beaker so that the large end is in the colored water.
3. Transfer a stalk of *celery* of approximately the same length and width (but that was kept in the air after being cut) into the "dry" beaker so that the large end is in the colored water.
4. With scissors, cut off the top end of each stalk, leaving about 10 centimeters total length.
5. Time how long it takes for the red-colored water to reach the top of each stalk, and record these data in Table 14.2.
6. In which celery stalk was the water column broken? _____

   _____

   Use this information to write a conclusion in Table 14.2.
7. Make a thin cross-sectional wet mount of the stalk in the "wet" beaker. Observe this slide under the microscope. *What type of tissue has been stained by the dye?*

**Figure 14.5** **Xylem structure.**
Xylem contains two types of conducting cells: tracheids and vessel elements. Tracheids have pitted walls, but vessel elements are larger and form a continuous pipeline from the roots to the leaves.

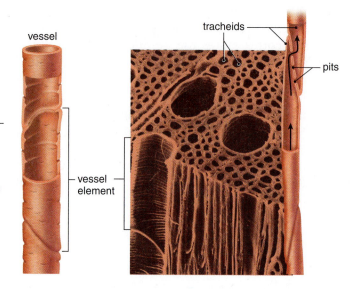

| Table 14.2 | Celery Stalk Experiment | |
|---|---|---|
| **Stalk** | **Speed of Dye (Minutes)** | **Conclusion** |
| Cut end placed in water prior to experiment | | |
| Cut end kept in air prior to experiment | | |

# 14.4  Organization of Stems

**Stems** are usually found aboveground and provide support for leaves and flowers. Stems that do not contain wood are called **herbaceous,** or nonwoody, stems. Herbaceous stems undergo primary growth but they do not undergo secondary growth. Activity of meristem in a terminal bud results in primary growth of a stem.

## Anatomy of Herbaceous Stems

Usually, monocots remain herbaceous throughout their lives. Some eudicots, such as those that live a single season, are also herbaceous. Other eudicots, such as trees, become woody as they mature.

### Observation: Anatomy of Eudicot and Monocot Herbaceous Stems

1. Observe a prepared slide of a eudicot stem such as *Helianthus* (sunflower). With the help of Figure 14.6a, identify

    **Epidermis:** the outermost protective layer.

    **Cortex:** may aid photosynthesis and store nutrients.

    **Vascular bundles:** (1) transports water and organic nutrients and (2) occurs in a ring pattern. Label the vascular bundle in Figure 14.6a.

    **Pith:** stores organic nutrients.

2. Observe a prepared slide of a monocot stem such as *Zea mays* (corn) (Fig. 14.6b). The vascular bundles are scattered.

3. State the main difference between the eudicot stem and the monocot stem. _____

    _____

**Figure 14.6  Nonwoody (herbaceous) stems.**
*a.* In eudicot stems, the vascular bundles are arranged in a ring around well-defined ground tissue called pith.
*b.* In monocot stems, the vascular bundles are scattered within the ground tissue.

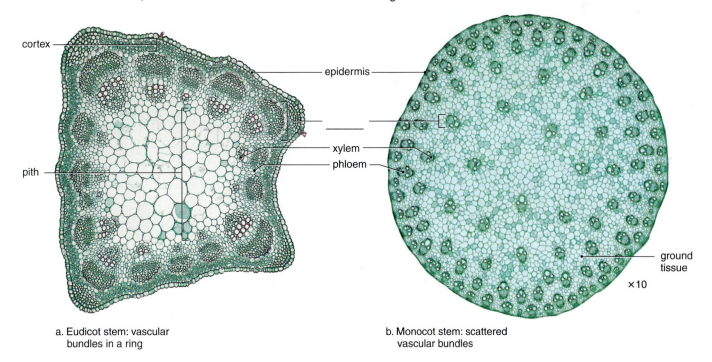

a. Eudicot stem: vascular bundles in a ring

b. Monocot stem: scattered vascular bundles

## Anatomy of Woody Stems

Woody stems undergo both primary growth (increase in length) and secondary growth (increase in girth). When *primary growth* occurs, **shoot apical meristem** within a terminal bud is active and **root apical meristem** within a root tip is active. When *secondary growth* occurs, vascular cambium is active. **Vascular cambium** is meristem tissue, which produces new xylem and phloem called **secondary xylem** and **phloem** each year. The build up of secondary xylem year after year is called **wood.**

### Observation: Anatomy of Winter Twig

1. A winter twig typically shows several years' past primary growth. Examine two examples of winter twigs (Fig. 14.7), and identify the **terminal bud** located at the tip of the twig. Primary growth of a stem originates here because a terminal bud contains shoot apical meristem.
2. Locate a **terminal bud scar.** These scars encircle the twig and indicate where the terminal bud was located in previous years. The distance between two adjacent terminal bud scars equals one year's primary growth.
3. Find a **leaf scar.** This is where a leaf was attached to the stem.
4. Identify a **node.** This is where a leaf scar is located.
5. Note the **vascular bundle scars.** Complete this sentence: Vascular bundle scars appear where the vascular bundles _____.
6. Locate an **axillary bud.** This is where new branch growth can occur because an axillary bud also contains meristem tissue.

**Figure 14.7** **External structure of a winter twig.**
Counting the terminal bud scars tell the age of a particular branch.

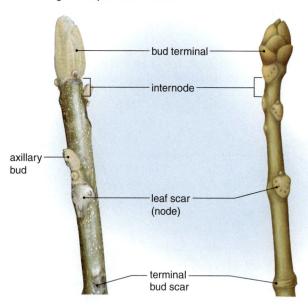

## Observation: Anatomy of Woody Stem

1. Examine a prepared slide of a cross section of a woody stem such as *Tilia* (tulip tree) (Fig. 14.8), and identify the **bark** (the dark outer area), which contains **cork,** a protective outer layer; **cortex,** which stores nutrients; and **phloem,** which transports organic nutrients.
2. Locate the **vascular cambium** at the inner edge of the bark, between the bark and the wood.
3. Find the **wood,** which contains annual rings. An **annual ring** is the amount of xylem added to the plant during one growing season. Rings appear to be present because spring wood has large xylem vessels and looks light in color, while summer wood has much smaller vessels and appears much darker. How old is the stem you are observing? _____ Are all the rings the same width? _____
4. Identify the **pith,** a ground tissue that stores organic nutrients and may disappear.
5. Locate **rays,** groups of small, almost cuboidal cells that laterally extend out from the pith.

**Figure 14.8** **Woody eudicot stem cross section.**
Because xylem builds up year after year, it is possible to count the annual rings to determine the age of a tree. This tree is three years old.

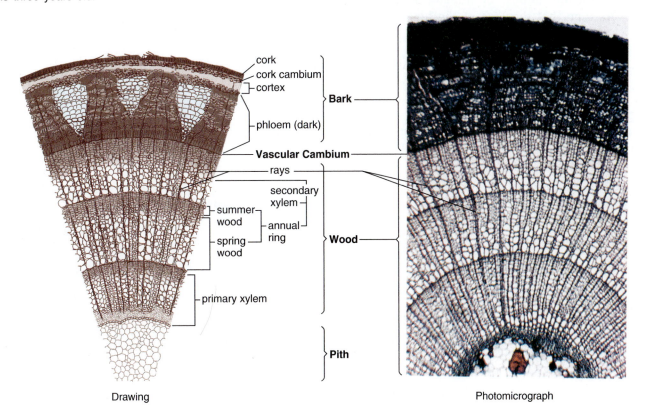

cork
cork cambium
cortex
Bark
phloem (dark)
Vascular Cambium
rays
secondary xylem
summer wood
annual ring
spring wood
Wood
primary xylem
Pith
Drawing

Photomicrograph

# 14.5 Organization of Leaves

Leaves are generally broad and quite thin to better capture solar energy. Carbon dioxide enters a leaf at openings called **stomata** (sing., stoma), and water enters by way of **leaf veins,** extensions of the vascular bundles from the stem.

## Observation: Stomata

1. Obtain a *leaf* from a plant designated by your instructor, and put a drop of *distilled water* on a slide.
2. Using the technique shown in Figure 14.9a, obtain a strip of outer tissue from a leaf. This tissue is epidermis.
3. Put the tissue in the drop of water on the slide, outer side up. Add a coverslip, and examine it microscopically, using both low power and high power.
4. Observe a stoma and the two guard cells that regulate the opening and closing of the stoma (Figure 14.9b).

    What gas enters a leaf at the stomata? _____

5. Count the number of stomata in the high-power field of view: _____.
6. Assume that the area of the high-power field is 0.10 mm. Divide the number of stomata by this area to determine the number of stomata in 1 square millimeter: _____.
7. Does your leaf contain a large number of stomata per square millimeter? _____

## Figure 14.9 Stomata.

*a.* Method of obtaining a strip of epidermis from the underside of a leaf. *b.* Photo to help you identify stomata under the microscope. Each stoma is bordered by two guard cells.

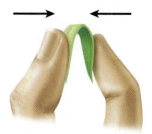

1. Pinch leaf.

2. Remove epidermis with forceps.

a.

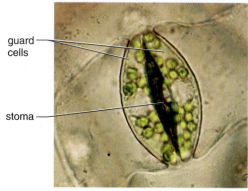

guard cells

stoma

b.

1. Obtain a slide of a leaf, such as *Liqustrum* (common privet) or *Syringa* (barley) in cross section.
2. With the help of Figure 14.10, identify

   **Cuticle:** the outermost layer that protects the leaf and prevents water loss.

   **Upper and lower epidermis:** a single layer of protective cells at the upper and lower surfaces.

   **Leaf vein:** transports water and organic nutrients. What tissues do a leaf vein contain?

   _____ and _____

   **Palisade mesophyll:** located near the upper epidermis. *Label the palisade mesophyll in Figure 14.10.*

   **Spongy mesophyll:** located near the lower epidermis. *Label the spongy mesophyll in Figure 14.10.*

   Which type of mesophyll has chloroplasts? _____

   Which type of mesophyll carries on photosynthesis? _____

   Which type of mesophyll has air spaces that facilitate exchange of gases? _____

   **Stomata:** the openings you examined in the previous observation.

## Figure 14.10 Leaf anatomy.

Complete the labels as directed in the Observation on this page.

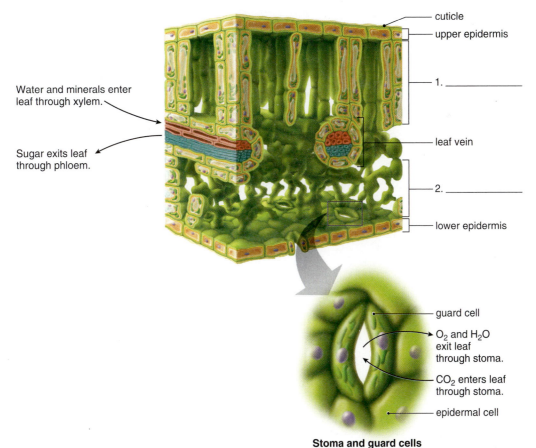

Water and minerals enter leaf through xylem.

Sugar exits leaf through phloem.

cuticle
upper epidermis
1. _____
leaf vein
2. _____
lower epidermis

guard cell
$O_2$ and $H_2O$ exit leaf through stoma.
$CO_2$ enters leaf through stoma.
epidermal cell

**Stoma and guard cells**

_____ 1. The leaves attach to what portion of a stem?

_____ 2. What type of venation do monocot leaves have?

_____ 3. State the function of the structures called stomata that are present in leaf epidermis.

_____ 4. What are the cells between the upper and lower epidermis of the leaf called?

_____ 5. Identify the slide if stomata are present.

_____ 6. What is the pattern for vascular bundle distribution in a monocot stem?

_____ 7. In woody stems, the bark is divided from the wood by what tissue?

_____ 8. Identify the slide if there are annual rings present.

_____ 9. Pith is most likely present in a _____.

_____ 10. What zone follows the zone of cell division in a root tip?

_____ 11. Lateral roots develop from which cell layer?

_____ 12. Epidermis is modified in a root by the addition of _____.

_____ 13. Identify the slide if the center tissue has several "arms" that extend like the spokes of a wheel.

_____ 14. What type of tissue transports materials in a plant?

## Thought Questions

15. If only slides of root and stem were available to you, how could you identify a plant as an herbaceous eudicot?

16. Contrast the manner in which water reaches the inside of a leaf with the manner in which carbon dioxide reaches the inside of a leaf.

# 15

# Animal Diversity

## Learning Objectives

**15.1 Classification of Animals**
- List the anatomical features that serve as a basis for the classification of animals.
- State the general characteristics of animals that have a coelom.

**15.2 Invertebrates**
- Describe the general characteristics of molluscs.
- Identify and/or locate the external and internal structures of a clam.
- Explain how clams and squids are adapted to their way of life.
- Describe the general characteristics of arthropods.
- Identify and/or locate external structures of the crayfish and the grasshopper.
- Contrast the external anatomy of the crayfish and the grasshopper, indicating how each is adapted to its way of life.
- Contrast complete and incomplete metamorphosis.

**15.3 Vertebrates**
- Name the vertebrate classes, giving an example of an animal in each class.
- Trace the path of air, food, and urine in a frog, and explain the term urogenital system.

**15.4 Comparison of Invertebrates with Vertebrates**
- Compare the crayfish anatomy with vertebrate anatomy and state similarities and dissimilarities.

## Introduction

This laboratory concerns the animal kingdom. Animals are all multicellular and have varying degrees of motility. They are heterotrophic and most digest their food in a digestive cavity.

One of the themes of today's laboratory will be the similarities and differences between the animals studied. The molluscs (e.g., clam) and arthropods (e.g., crayfish and grasshopper) are invertebrates, while the amphibians (e.g., frog) are vertebrates. Invertebrates lack a backbone of vertebrae that is present in vertebrates. All of the animals in today's laboratory have true tissues, bilateral symmetry, tube-within-a-tube body plan, and a coelom or body cavity. They differ in that the molluscs are nonsegmented, while the arthropods and vertebrates are segmented animals with jointed appendages. Jointed appendages are particularly adapted for locomotion on land. The insects are arthropods that are plentiful on land. Among the vertebrates, the amphibians live on land at least part of their lives, while the reptiles, birds, and mammals are primarily land animals (Fig. 15.1).

**Figure 15.1  Animal diversity.**
The chameleon, a reptile, and the fly, an arthropod, are both very well adapted to living on land. The skin of a reptile and the external skeleton of an arthropod resist drying out. They both breathe by taking in air and both have a suitable means of locomotion on land.

## Figure 15.2  Evolution of animals.

All animals are believed to be descended from a type of protist; however, sponges may have evolved from a protist separately from the rest of the animals.

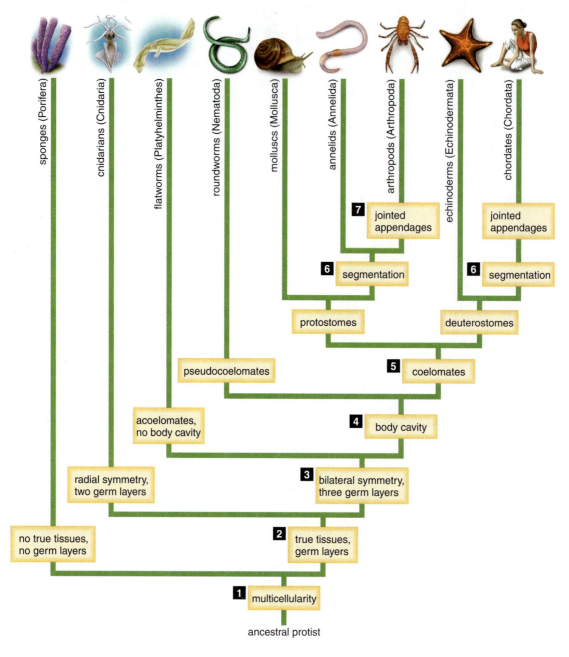

# 15.1 Classification of Animals

The evolutionary tree of animals is given in Figure 15.2 and the anatomical features that help classify animals is given in Table 15.1. Using the evolutionary tree, circle the features possessed by all the animals that will be studied today.

| Table 15.1 | Anatomical Features for Animal Classification |
| --- | --- |
| **Level of Organization** | |
| Cellular | Cells not organized into tissues |
| Tissues | Cells organized into tissues |
| Organs | Tissues organized into organs |
| Organ systems | Organs organized into systems |
| **Germ Layers** | |
| None | |
| Two | Two germ layers: the ectoderm (outside) and the endoderm (inside) |
| Three | Three germ layers: the ectoderm, the endoderm, and the mesoderm (middle); presence of mesoderm allows for development of various internal organs |
| **Symmetry** | |
| None | |
| Radial | Any longitudinal cut through the midpoint yields equal halves; this design allows animals to reach out in all directions |
| Bilateral | Only one longitudinal cut through the midpoint yields equal halves; such animals often have well-developed head regions (cephalization) |
| **Body Plan** | |
| None | |
| Sac plan | Mouth used for intake of nutrient molecules and exit of waste molecules |
| Tube-within-a-tube | Separate openings (mouth and anus) for food intake and waste exit; allows for specialization of parts along the digestive canal |
| **Coelom** | |
| Acoelomate | Have no coelom (fluid-filled extracellular spaces) |
| Pseudocoelomate | Have false coelom; coelom incompletely lined with mesoderm |
| Coelomate | Have true coelom, a fluid-filled body cavity completely lined with mesoderm |
| **Segmentation** | |
| Nonsegmented | No segmentation (repeating parts) |
| Segmented | A repeating series of parts from anterior to posterior; allows for specialization of some sections for different functions |

## 15.2 Invertebrates

All of the phyla in Figure 15.2 contain invertebrates. The vertebrates only occur in phylum chordata. Most types of invertebrates are adapted to living in the sea; the insects being the major exception to this statement. We will have an opportunity to contrast adaptations to the land environment with adaptations to the aquatic environment.

### Molluscs

Most **molluscs** are marine, but there are also some freshwater and terrestrial molluscs (Fig. 15.2). All molluscs have a three-part body consisting of a ventral, muscular **foot** specialized for various means of locomotion; a **visceral mass** that includes the internal organs; and a **mantle,** a thin tissue that encloses the visceral mass and may secrete a shell.

### Observation: Diversity of Molluscs

Most molluscs belong to one of four classes. Class Polyplacophora contains grazing marine herbivores, such as chitons, with a body flattened dorsoventrally covered by a shell consisting of eight plates (Fig. 15.3). Class Bivalvia contains marine and freshwater sessile filter feeders, such as clams, with a body enclosed by a shell consisting of two valves. These animals have a hatchet-shaped foot but no head or radula. Class Cephalopoda contains marine active predators, such as squids. The shell may be reduced or absent; the head which is anterior to an elongated visceral mass bears a circle of tentacles and arms. The circulatory system is efficient, and the well-developed nervous system is accompanied by cephalization. Class Gastropoda contains marine, freshwater, and often terrestrial herbivores, such as snails, with a coiled shell that distorts body symmetry. The head has tentacles.

1. Examine mollusc specimens, and complete the first two columns of Table 15.2.
2. Examine the foot, and complete the third column of Table 15.2. Some molluscs have a broad, flat foot, others a hatchet-shaped foot; in still others the foot has become tentacles and arms that assist in the capture of food.
3. Indicate in the last column if cephalization is present or not present.

**Figure 15.3   Three classes of molluscs.**
Top row: Snails (*left*) and nudibranchs (*right*) are gastropods. Middle row: Octopuses (*left*) and nautiluses (*right*) are cephalopods. Bottom row: Scallops (*left*) and mussels (*right*) are bivalves.

| Table 15.2 | Molluscan Diversity | | |
|---|---|---|---|
| Common Name of Specimen | Class | Description of Foot | Cephalization (Yes or No) |
| | | | |
| | | | |
| | | | |
| | | | |

## Anatomy of Clam

Clams are bivalved because they have right and left shells secreted by the mantle. Clams have no head, and they burrow in sand by extending a "hatchet" foot between the valves. Clams are filter feeders and feed on debris that enters the mantle cavity. Clams have an open circulatory system; the blood leaves the heart and enters sinuses (cavities) by way of anterior and posterior aortas. There are many different types of clams. The one examined here is the freshwater clam *Venus*.

### Observation: Anatomy of Clam

**External Anatomy**

1. Examine the external shell (Fig. 15.4) of a preserved clam (*Venus*). The shell is an **exoskeleton.**
2. Find the posterior and anterior ends. The more pointed end of the **valves** (the halves of the shell) is the posterior end.
3. Determine the clam's dorsal and ventral regions. The valves are hinged together dorsally.
4. What is the function of a heavy shell? _____

**Figure 15.4** **External view of clam shell.**

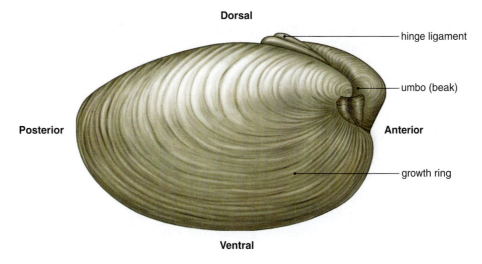

## Internal Anatomy

1. Place the clam in the dissecting pan, with the **hinge ligament** and **umbo** (blunt dorsal protrusion) down. Carefully separate the **mantle** from the right valve by inserting a scalpel into the slight opening of the valves. What is a mantle? _____

_____

2. Insert the scalpel between the mantle and the valve you just loosened.
3. The **adductor muscles** hold the valves together. Cut the adductor muscles at the anterior and posterior ends by pressing the scalpel toward the dissecting pan. After these muscles are cut, the valve can be carefully lifted away. What is the advantage of powerful adductor muscles? _____

_____

4. Examine the inside of the valve you removed. Note the concentric lines of growth on the outside, the hinge teeth that interlock with the other valve, the adductor muscle scars, and the mantle line. The inner layer of the shell is mother-of-pearl.
5. Examine the rest of the clam (Fig. 15.5) attached to the other valve. Notice the mantle, which lies over the visceral mass and foot, and also the adductor muscles.
6. Bring the two halves of the mantle together. Explain the term *mantle cavity.* _____

_____

7. Identify the **incurrent** (more ventral) and **excurrent siphons** at the posterior end (Fig. 15.5). Explain how water enters and exits the mantle cavity. _____

_____

8. Cut away the free-hanging portion of the mantle to expose the **gills.** Does the clam have a respiratory organ? _____

What type of respiratory organ? _____
9. A mucous layer on the gills entraps food particles brought into the mantle cavity, and the cilia on the gills convey these food particles to the mouth. Why is the clam called a filter feeder?

_____

_____

10. The nervous system is composed of three pairs of ganglia (located anteriorly, posteriorly, and in the foot), all connected by nerves. The clam does not have a brain. A ganglion contains a limited number of neurons, whereas a brain is a large collection of neurons in a definite head region.
11. Identify the **foot,** a tough, muscular organ for locomotion, and the **visceral mass,** which lies above the foot and is soft and plump. The visceral mass contains the digestive and reproductive organs.
12. Identify the **labial palps** that channel food into the open mouth.
13. Identify the **anus,** which discharges into the excurrent siphon.
14. Find the **intestine** by its dark contents. Trace the intestine forward until it passes into a sac, the clam's only evidence of a coelom.
15. Locate the **pericardial sac (pericardium)** that contains the heart. The intestine passes through the heart. The heart pumps blood into the aortas, which deliver it to blood sinuses in the tissues.

A clam has an **open circulatory system.** Explain. _____

_____

16. Cut the visceral mass and the foot into exact left and right halves, and examine the cut surfaces. Identify the greenish-brown digestive glands; the stomach, embedded in the digestive glands; and the intestine, which winds about in the visceral mass. Reproductive organs are also present.

## Figure 15.5 Anatomy of a bivalve.

The mantle has been removed to reveal the internal organs. (a) Drawing. (b) Dissected specimen.

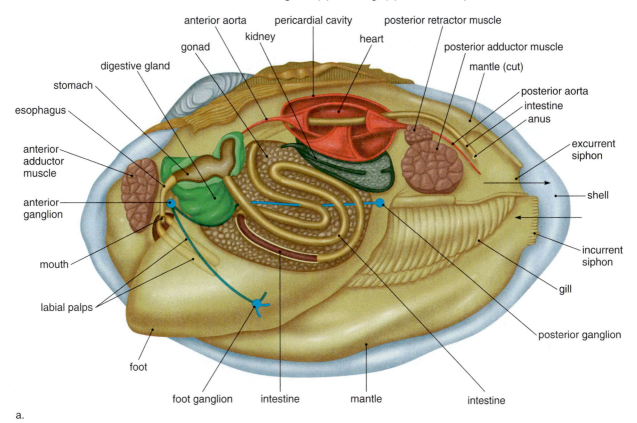

a.

b.

## Anatomy of Squid

Squids are cephalopods (means head-foot) because they have a well-defined head. The head contains a brain and bears sense organs. The squid moves quickly by jet propulsion of water, which enters the mantle cavity by way of a space that circles the head. When the cavity is closed off, water exits by means of a funnel. Then the squid moves rapidly in the opposite direction.

The squid seizes fish with its tentacles; the mouth has a pair of powerful, beaklike jaws and a **radula,** a filelike organ containing rows of teeth. The squid has a closed circulatory system composed of vessels and three hearts, one of which pumps blood to all the internal organs, while the other two pump blood to the gills located in the mantle cavity.

### Observation: Anatomy of Squid

1. Examine a preserved squid.
2. Refer to Figure 15.6 for help in identifying the mouth (defined by beaklike jaws and containing a radula) and the tentacles and arms, which encircle the mouth.
3. Locate the head with its sense organs, notably the large, well-developed eye.
4. Find the funnel, where water exits from the mantle cavity, causing the squid to move backward.
5. If the squid has been dissected, note the heart, gills, and blood vessels.

### Clam Anatomy Compared with Squid Anatomy

1. Compare clam anatomy with squid anatomy by completing Table 15.3.
2. Explain how both clams and squids are adapted to their way of life.

_____

_____

_____

| Table 15.3 | Comparison of Clam and Squid | |
|---|---|---|
| | Clam | Squid |
| Feeding mode | | |
| Skeleton | | |
| Circulation | | |
| Cephalization | | |
| Locomotion | | |
| Nervous system | | |

## Figure 15.6 Anatomy of a squid.

The squid is an active predator and lacks the external shell of a clam. It captures fish with its tentacles and bites off pieces with its jaws. A strong contraction of the mantle forces water (arrows) out the funnel, resulting in "jet propulsion." (*a*) Drawing. (*b*) Dissected specimen.

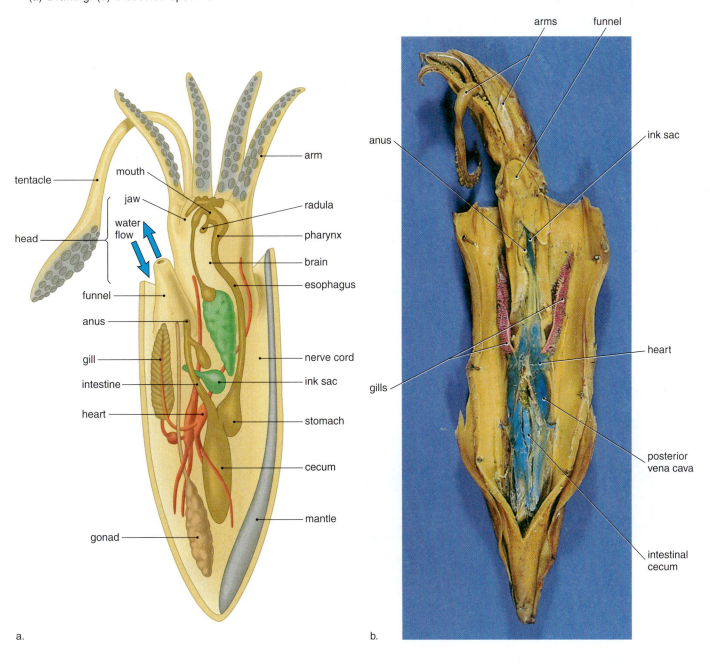

a.

b.

## Arthropods

**Arthropods** have paired, jointed appendages and a hard exoskeleton that contains chitin. The chitinous exoskeleton consists of hardened plates separated by thin, membranous areas that allow movement of the body segments and appendages. Arthropods are segmented like the annelids, but specialization of segments has occurred. Explain. _____

_____

The arthropods are divided into three groups. In one group are the spiders, scorpions, and horseshoe crabs; in another are the millipedes, centipedes, and insects; and the third group consists of the crustaceans (e.g., crabs) and barnacles.

### Observation: Diversity of Arthropods

1. Examine various specimens of arthropods (Fig. 15.7), and complete Table 15.4.
2. In the last column, note the number and type of appendages attached to the thorax and/or abdomen.

### Figure 15.7 Crustacean diversity.

(*a*) A copepod uses its long antennae for floating and its feathery maxillae for filter feeding. Shrimp (*b*) and crabs (*c*) are decapods—they have five pairs of walking legs. Shrimp resemble crayfish more closely than they do crabs, which have a reduced abdomen. Marine shrimp feed on codepods. Barnacles have no abdomen and a reduced head; the thoracic legs project through a shell to filter feed. Barnacles often live on humanmade objects such as ships, buoys, and cables. (*d*) The gooseneck barnacle attaches to an object by a long stalk.

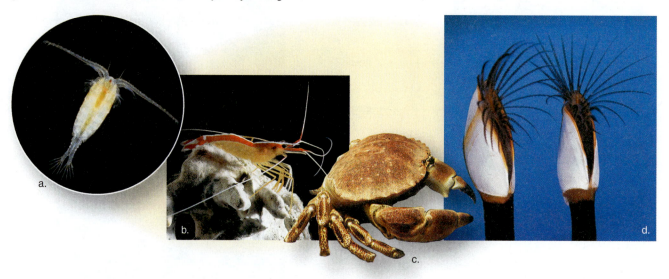

| Table 15.4 | Arthropod Diversity | |
|---|---|---|
| **Common Name of Specimen** | **Subphylum** | **Appendages (Attached to Body)** |
| | | |
| | | |
| | | |
| | | |

## Anatomy of Crayfish

Crayfish belong to a group of arthropods called crustaceans (Fig. 15.7). Crayfish are adapted to an aquatic existence. They are known to be scavengers, but they also prey on other invertebrates. The mouth is surrounded by appendages modified for feeding, and there is a well-developed digestive tract. Dorsal, anterior, and posterior arteries carry hemolymph (blood plus lymph) to tissue spaces (hemocoel) and sinuses. In contrast to vertebrates, there is a ventral solid nerve cord.

### Observation: Anatomy of Crayfish

#### External Anatomy

1. Obtain a preserved crayfish, and place it in a dissecting pan.
2. Identify the chitinous **exoskeleton.** With the help of Figure 15.8, identify the head, thorax, and abdomen. Together, the head and thorax are called the **cephalothorax;** the cephalothorax is

### Figure 15.8   Anatomy of a crayfish.
*a.* Drawing of external anatomy (male). *b.* Dissection of internal anatomy (female).

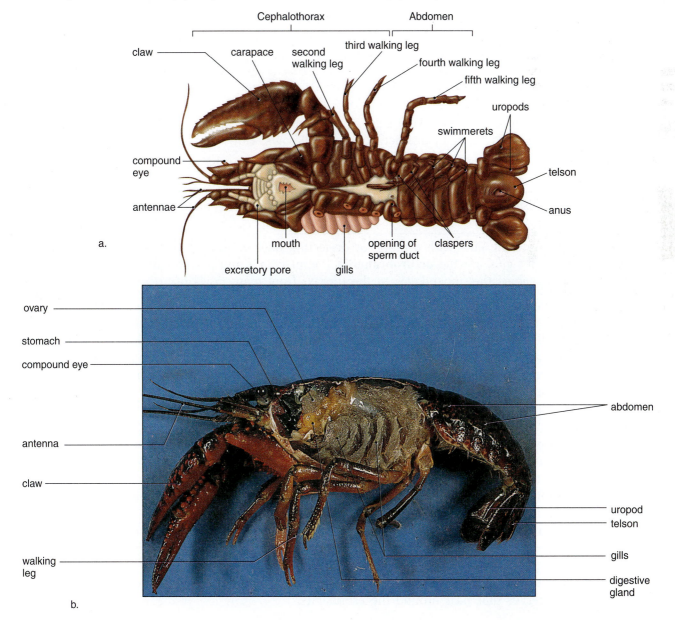

Animal Diversity   Laboratory 15   **191**

covered by the **carapace.** Has specialization of segments occurred? _____ Explain. _____

3. Find the **antennae,** which project from the head. At the base of each antenna, locate a small, raised nipple containing an opening for the **green glands,** the organs of excretion. Crayfish excrete a liquid nitrogenous waste.

4. Locate the **compound eyes,** composed of many individual units for sight. Do crayfish demonstrate cephalization? _____ Explain. _____

5. Identify the six pairs of appendages around the mouth for handling food.

6. Find the five pairs of walking legs attached to the cephalothorax. The most anterior pair is modified as pincerlike claws.

7. Locate the five pairs of **swimmerets** on the abdomen. In the male, the anterior two pairs are stiffened and folded forward. They are claspers that aid in the transfer of sperm during mating.

8. In the female, identify the **seminal receptacles,** a swelling located between the bases of the third and fourth pairs of walking legs. Sperm from the male are deposited in the seminal receptacles. In the male, identify the opening of the sperm duct located at the base of the fifth walking leg.

   What sex is your specimen? _____

9. Examine the opposite sex also.

10. Find the last abdominal segment, which bears a pair of broad, fan-shaped **uropods** that, together with a terminal extension of the body, form a tail. Has specialization of appendages occurred? _____ Explain. _____

### Internal Anatomy

1. Place the crayfish in the dissecting pan.

2. Cut away the lateral surface of the carapace with scissors to expose the **gills** (Fig. 15.8b). Observe that the gills occur in distinct, longitudinal rows. How many rows of gills are there in your specimen? _____ The outer row of gills is attached to the base of certain appendages. Which ones? _____
These outer gills are the **podobranchia** ("foot gills"). How many podobranchia do you find in your specimen? _____

3. Carefully separate the gills with a probe or dissecting needle, and locate the inner row(s) of gills. These inner gills are the **arthrobranchia** ("joint gills") and are attached to the chitinous membrane that joins the appendages to the thorax. How many rows of arthrobranchia do you find in the specimen? _____

4. Remove a gill with your scissors by cutting it free near its point of attachment, and place it in a watch glass filled with water. Observe the numerous gill filaments arranged along a central axis.

5. Carefully cut away the dorsal surface of the carapace with scissors and a scalpel. The epidermis that adheres to the exoskeleton secretes the exoskeleton. Remove any epidermis adhering to the internal organs.

6. Identify the diamond-shaped heart lying in the middorsal region. A crayfish has an open circulatory system. Carefully remove the heart.

7. Locate the **gonads** anterior to the heart in both the male and female. The gonads are tubular structures bilaterally arranged in front of the heart and continuing behind it as a single mass. In the male, the testes are highly coiled, white tubes.

8. Find the **mouth;** the short, tubular **esophagus;** and the two-part **stomach,** with the attached **digestive gland,** that precedes the intestine.

9. Identify the **green glands,** two excretory structures just anterior to the stomach, on the ventral segment wall.

10. Remove the thoracic contents previously identified.

11. Identify the **brain** in front of the esophagus. The brain is connected to the ventral nerve cord by a pair of nerves that pass around the esophagus.

12. Remove the animal's entire digestive tract, and float it in water. Observe the various parts, especially the connections of the digestive gland to the stomach.

13. Cut through the stomach, and notice in the anterior region of the stomach wall the heavy, toothlike projections, called the **gastric mill,** that grind up food. Do you see any grinding stones ingested by the crayfish? _____

If possible, identify what your specimen had been eating. _____

## Anatomy of Grasshopper

The grasshopper is an **insect** (Fig. 15.9). Insects are adapted to life on land. In insects with wings, such as the grasshopper, wings are attached to the thorax. Respiration is by a highly branched internal system of tubes, called tracheae.

### Observation: Anatomy of Grasshopper

**External Anatomy**

1. Obtain a preserved grasshopper (*Romalea*), and study its external anatomy with the help of Figure 15.10. Identify the head, thorax, and abdomen.

2. The grasshopper's **thorax** consists of three fused segments: the large anterior **prothorax,** the middle **mesothorax,** and the hind **metathorax.** Identify the first pair of legs attached to the prothorax. Then find the second pair of legs and the outer pair of straight, leathery **forewings** attached to the mesothorax. Finally, locate the third pair of legs and the inner, membranous **hind wings** attached to the metathorax. Each leg consists of five segments. The hind leg is well developed and used for jumping. How many pairs of legs are there? _____

**Figure 15.9  Insect diversity.**

Housefly

Green lacewing

Walking stick

Scarlet and green leafhopper

Tortoiseshell scale

Flea

Snout beetle

Dragonfly

Luna moth

Honeybee

Lubber grasshopper

## Figure 15.10 External anatomy of a female grasshopper, *Romalea.*

a. The legs and wings are attached to the thorax. b. The head has mouthparts of various types.

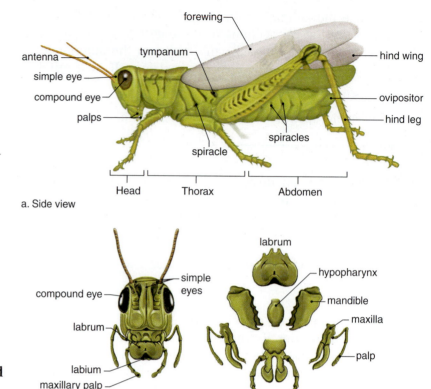

a. Side view

3. Is locomotion in the grasshopper adapted to land?

_____

Explain. _____

_____

_____

4. Use a hand lens or dissecting microscope to examine the grasshopper's special sense organs of the head. Identify the **antennae** (a pair of long, jointed feelers); the **compound eyes;** and the three, dotlike **simple eyes.**

5. Remove the **mouthparts** by grasping them with forceps and pulling them out. Arrange them in order on an index card, and compare them with Figure 15.10b. These mouthparts are used for chewing and are quite different from those of a piercing and sucking insect.

b. Head and mouthparts

6. Identify the **tympana** (sing., **tympanum**), one on each side of the first abdominal segment (Fig. 15.10a). The grasshopper detects sound vibrations with these membranes.

7. Locate the **spiracles,** along the sides of the abdominal segments. These openings allow air to enter the tracheae, which constitute the respiratory system.

8. Find the **ovipositors** (Fig. 15.11a), four curved and pointed processes projecting from the hind end of the female. These are used to dig a hole in which eggs are laid. The male has **claspers** that are used during copulation (Fig. 15.11b).

## Figure 15.11 Grasshopper genitalia.

(a) Females have an ovipositor, and (b) males have claspers.

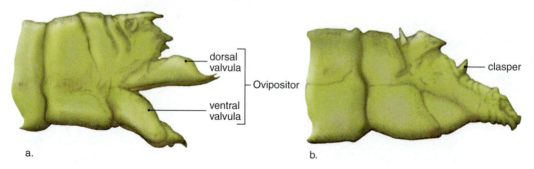

a.

b.

**Internal Anatomy**

1. Detach the wings and legs of the grasshopper identified in number 2, page 193. Then turn the organism on its side, and use scissors to carefully cut through the exoskeleton (dorsal to the spiracles) along the full length (from the head to the posterior end) of the animal. Repeat this procedure on the other side.
2. Cut crosswise behind the head so that you can remove a strip of the exoskeleton. If necessary, reach in with a probe to loosen the muscle attachments and membranes.
3. Pin the insect to the dissecting pan, dorsal side up. Cover the specimen with water to keep the tissues moist.
4. Identify the heart (Fig. 15.12) and aorta just beneath the portion of exoskeleton you removed. A grasshopper has an open circulatory system. Remove the heart and adjacent tissues.
5. Locate the **fat body,** a yellowish fatty tissue that covers the internal organs. Carefully remove it.
6. Find the **tracheae,** the respiratory system of insects. Using the dissecting microscope, look for glistening white tubules, which deliver air to the muscles.
7. Identify the reproductive organs that lie on either side of the digestive tract in the abdomen. If your specimen is a male, look for the testis, a coiled, elongated cord containing many tubules. If your specimen is a female, look for the ovary, essentially a collection of parallel, tapering tubules containing cigar-shaped eggs.
8. Locate the digestive tract and, in sequence, the **crop,** a large pouch for storing food (a grasshopper eats grasses); the **gastric ceca,** digestive glands attached to the stomach; the stomach and the intestine, which continues to the anus; and **Malpighian tubules,** excretory organs attached to the intestine. Insects secrete a dry solid nitrogenous waste. Is this an

    adaptation to life on land? _____ Explain. _____

    _____
9. Work the digestive tract free, and move it to one side. Now identify the **salivary glands** that extend into the thoracic cavity.
10. Remove the internal organs. Now identify the ventral **nerve cord,** thickened at intervals by ganglia.
11. Remove one side of the exoskeleton covering the head. Identify the brain, anterior to the esophagus.

**Figure 15.12  Internal anatomy of a female grasshopper.**
The digestive system of a grasshopper shows specialization of parts.

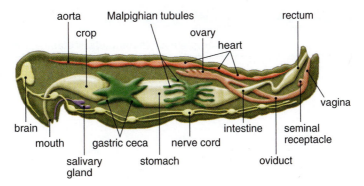

**Conclusion**

Compare the adaptations of a crayfish with those of a grasshopper by completing Table 15.5. Put a star beside each item that indicates an adaptation to life in the water (crayfish) and to life on land (grasshopper). How many did you identify? _____ Check with your instructor to see if you identified the maximum number of adaptations.

| Table 15.5 | Comparison of Crayfish and Grasshopper | |
| --- | --- | --- |
| | Crayfish | Grasshopper |
| Locomotion | | |
| Respiration | | |
| Nervous system | | |
| Reproductive features | | |
| Sense organs | | |

## Insect Metamorphosis

**Metamorphosis** means a change, usually a drastic one, in form and shape. Some insects undergo what is called *complete metamorphosis,* in which case they have three stages of development: the larval stages, the pupa stage, and finally the adult stage. Metamorphosis occurs during the pupa stage, when the animal is enclosed within a hard covering. The animals best known for metamorphosis are the butterfly and the moth, whose larval stage is called a caterpillar and whose pupa stage is the cocoon; the adult is the butterfly or moth (Fig. 15.13a). Grasshoppers undergo *incomplete metamorphosis,* a gradual change in form rather than a drastic change. The immature stages of the grasshopper are called nymphs rather than larvae, and they are recognizable as grasshoppers even though they differ somewhat in shape and form (Fig. 15.13b).

If available, examine life cycle displays or plastomounts that illustrate complete and incomplete metamorphosis.

### Observation: Insect Metamorphosis

Observe any insects available, and state in Table 15.6 whether they have complete metamorphosis or incomplete metamorphosis.

| Table 15.6 | Insect Metamorphosis |
| --- | --- |
| **Common Name of Specimen** | **Complete or Incomplete Metamorphosis** |
| | |
| | |
| | |
| | |
| | |
| | |

## Figure 15.13 Insect metamorphosis.

During (*a*) complete metamorphosis, a series of larvae leads to pupation. The adult hatches out of the pupa. During (*b*) incomplete metamorphosis, a series of nymphs leads to a full-grown grasshopper.

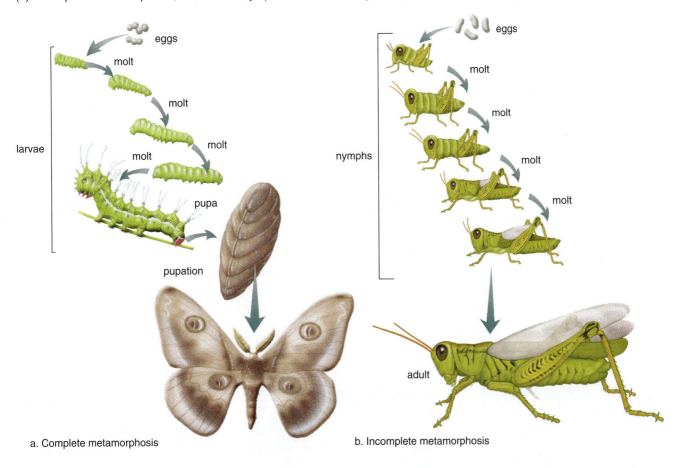

a. Complete metamorphosis

b. Incomplete metamorphosis

## Conclusions

- With reference to Figure 15.13, what stage is missing when an insect does not have complete metamorphosis? _____ What happens during this stage? _____
- What form, the larvae or the adult, disperses offspring in flying insects? _____ How is this a benefit? _____
- In insects that undergo complete metamorphosis, the larvae and the adults use different food sources and habitats. Why might this be a benefit? _____
- With reference to insects that undergo incomplete metamorphosis, which form, the nymphs or the adult, have better developed wings? _____ What is the benefit of wings to an insect?

_____

_____

# 15.3 Vertebrates

Vertebrates are a type of **chordate** (see Fig. 15.2). Despite their great diversity, all chordates at some time in their life history have four characteristics:

1. A **notochord**—a dorsal supporting rod extending the length of the body. The notochord is replaced during development by a vertebral column in the vertebrates.
2. A **dorsal tubular nerve cord**—In vertebrates, the nerve cord, more often called the **spinal cord,** is protected by the vertebrae.
3. **Pharyngeal pouches**—which become functioning **gills** in the invertebrate chordates, the fishes, and amphibian larvae. In terrestrial animals, the pouches are modified for various other functions.
4. A **post-anal tail**—as an embryo if not as an adult; a tail that extends beyond the anus.

## Diversity of Vertebrates

There are both aquatic and terrestrial vertebrates (as shown in Figure 15.14 and the classification table on page 205). Fishes are aquatic; adult amphibians may be terrestrial, but most must return to an aquatic habitat to reproduce; and reptiles are terrestrial, although some reptiles, such as sea turtles, are secondarily adapted to life in the water. Mammals are adapted to a wide variety of habitats, including air (e.g., bats), sea (e.g., whales), and land (e.g., humans).

**Figure 15.14** **Vertebrate classes.**
*a.* Class Chondrichthyes contains the cartilaginous fishes. *b.* Class Osteichthyes contains the bony fishes. *c.* Class Amphibia contains the frogs and salamanders. *d.* Class Reptilia contains the turtles, lizards, snakes, crocodiles, and alligators. *e.* Class Aves contains the birds. *f.* Class Mammalia contains the mammals.

a. Class Chondrichthyes: blue shark, *Prionace glauca*

b. Class Osteichthyes: blueback butterflyfish, *Chaetodon plebius*

c. Class Amphibia: northern leopard frog, *Rana pipiens*

d. Class Reptilia: Pearl River redbelly turtle, *Pseudemys* sp.

e. Class Aves: scissor-tailed flycatcher, *Aves tyrannidae*

f. Class Mammalia: grey fox, *Urocyon cinereoargenteus*

## Anatomy of Frog

Frogs are amphibians, a group of animals in which metamorphosis occurs. Metamorphosis includes a change in structure, as when an aquatic tadpole becomes a frog with lungs and limbs (Fig. 15.15). Amphibians were the first vertebrates to be adapted to living on land; however, they typically return to the water to reproduce. *Place a check in the margin below for every adaptation to a land environment.*

### Observation: External Anatomy of Frog

1. Place a preserved frog (*Rana pipiens*) in a dissecting tray.
2. Identify the bulging eyes, which have nonmovable upper and lower lids but can be covered by a **nictitating membrane** that moistens the eye.
3. Locate the **tympanum** behind each eye (Fig. 15.15). What is the function of a tympanum? _____

   _____
4. Examine the external **nares** (sing., **naris,** or **nostril**) (Fig. 15.15). Insert a probe into an external naris, and observe that it protrudes from one of the paired, small openings, the internal

   nares, inside the mouth cavity. What is the function of the nares? _____
5. Identify the paired limbs. The bones of the forelimbs and hindlimbs are the same as in all tetrapods, in that the first bone articulates with a girdle and the limb ends in phalanges. The hind feet have five phalanges, and the forefeet have only four phalanges. Which pair of limbs

   is longest? _____

   How does a frog locomote on land? _____

   _____

   What is a frog's means of locomotion in the water? _____

   _____

**Figure 15.15** **External frog anatomy.**

### Mouth

1.  Open your frog's mouth very wide (Fig. 15.16), cutting the angles of the jaws if necessary.
2.  Identify the tongue attached to the lower jaw's anterior end.
3.  Find the **auditory (eustachian) tube** opening in the angle of the jaws. These tubes lead to the ears. Auditory tubes equalize air pressure in the ears.
4.  Examine the **maxillary teeth** located along the rim of the upper jaw. Another set of teeth—**vomerine teeth**—is present just behind the midportion of the upper jaw.
5.  Locate the **glottis,** a slit through which air passes into and out of the **trachea,** the short tube from glottis to lungs. What is the function of a glottis? _____

    _____
6.  Identify the **esophagus,** which lies dorsal and posterior to the glottis and leads to the stomach.

    If available, examine life cycle displays or plastomounts that illustrate complete and incomplete metamorphosis.

**Figure 15.16** **Mouth cavity of a frog.**
*a.* Drawing. *b.* Dissected specimen.

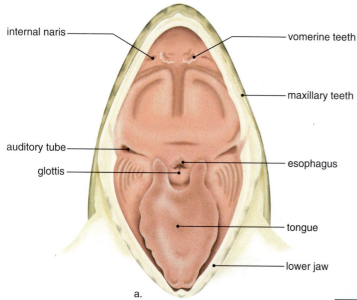

internal naris
vomerine teeth
maxillary teeth
auditory tube
glottis
esophagus
tongue
lower jaw

a.

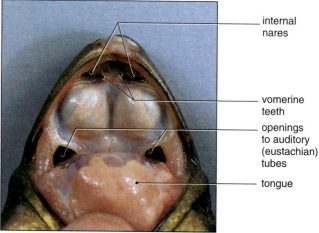

internal nares

vomerine teeth

openings to auditory (eustachian) tubes

tongue

b.

## Opening the Frog

1. Place the frog ventral side up in the dissecting pan. Lift the skin with forceps, and use scissors to make a large, circular cut to remove the skin from the abdominal region as close to the limbs as possible. Cut only skin, not muscle.
2. Now, remove the muscles by cutting through them in the same circular fashion. At the same time, cut through any bones you encounter. A vein, called the abdominal vein, will be slightly attached to the internal side of the muscles.
3. Identify the **coelom,** or body cavity.
4. If your frog is female, the abdominal cavity is likely to be filled by a pair of large, transparent **ovaries,** each containing hundreds of black and white eggs. Gently lift the left ovary with forceps, and find its place of attachment. Cut through the attachment, and remove the ovary in one piece.

## Respiratory System and Liver

1. Insert a probe into the glottis, and observe its passage into the trachea. Enlarge the glottis by making short cuts above and below it. When the glottis is spread open, you will see a fold on either side; these are the vocal cords used in croaking.
2. Identify the **lungs,** two small sacs on either side of the midline and partially hidden under the liver (Fig. 15.17). Trace the path of air from the external nares to the lungs. _____

_____

3. Locate the **liver,** the large, prominent, dark brown organ in the midventral portion of the trunk (Fig. 15.17). Between the right half and left half of the liver, find the **gallbladder.**

## Circulatory System

1. Lift the liver gently. Identify the **heart,** covered by a membranous covering (the **pericardium**). With forceps, lift the covering, and gently slit it open. The heart consists of a single, thick-walled **ventricle** and two (right and left) anterior, thin-walled **atria.**
2. Locate the three large veins that join beneath the heart to form the **sinus venosus.** (To lift the heart, you may have to snip the slender strand of tissue that connects the atria to the pericardium.) Blood from the sinus venosus enters the right atrium. The left atrium receives blood from the lungs.
3. Find the **conus arteriosus,** a single, wide arterial vessel leaving the ventricle and passing ventrally over the right atrium. Follow the conus arteriosus forward to where it divides into three branches on each side. The middle artery on each side is the **systemic artery,** which fuses behind the heart to become the **dorsal aorta.** The dorsal aorta transports blood through the body cavity and gives off many branches. The **posterior vena cava** begins between the two kidneys and returns blood to the sinus venosus. Which vessel lies above (dorsal to) the other? _____

_____

## Digestive Tract

1. Identify the **esophagus,** a very short connection between the mouth and the stomach. Lift the left liver lobe, and identify the stomach, whitish and J-shaped. The **stomach** connects with the esophagus anteriorly and with the small intestine posteriorly.
2. Find the **small intestine** and the **large intestine,** which enters the **cloaca** (see Figs. 15.18 and 15.19). The cloaca lies beneath the pubic bone and is a general receptacle for the intestine, the reproductive system, and the urinary system. It opens to the outside by way of the anus. Trace the path of food in the digestive tract from the mouth to the cloaca. _____

_____

## Accessory Glands

1. You identified the liver and gallbladder previously. Now try to find the **pancreas,** a yellowish organ near the stomach and intestine.

**Figure 15.17**  Internal organs of a female frog, ventral view.

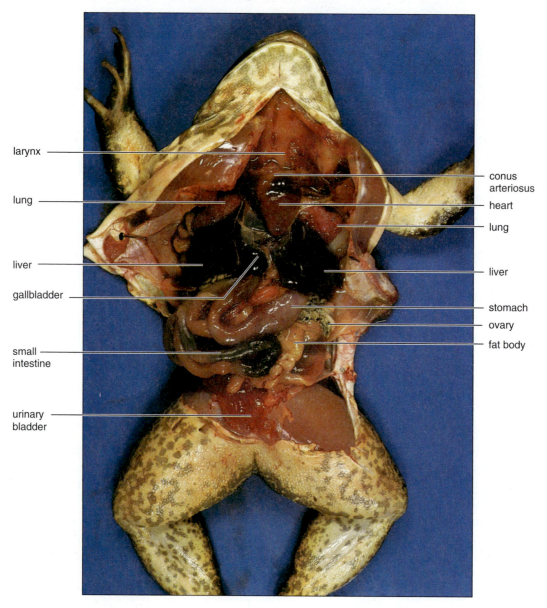

larynx

lung

liver

gallbladder

small
intestine

urinary
bladder

conus
arteriosus

heart

lung

liver

stomach

ovary

fat body

2.  Locate the **spleen,** a small, pea-shaped body near the stomach.

### Urogenital System

1.  Identify the **kidneys,** long, narrow organs lying against the dorsal body wall (Fig. 15.18).
2.  Locate the **testes** in a male frog (Fig. 15.18). Testes are yellow, oval organs attached to the anterior portions of the kidneys. Several small ducts, the **vasa efferentia,** carry sperm into kidney ducts that also carry urine from the kidneys. **Fat bodies,** which store fat, are attached to the testes.
3.  Locate the ovaries in a female frog. The ovaries are attached to the dorsal body wall (Fig. 15.19). Fat bodies are also attached to the ovaries. Highly coiled **oviducts** lead to the cloaca. The ostium (opening) of the oviduct is dorsal to the liver.
4.  Find the **mesonephric ducts**—thin, white tubes that carry urine from the kidney to the cloaca. In female frogs, you will have to remove the left ovary to see the mesonephric ducts.
5.  Locate the **cloaca.** You will need to split through the bones of the pelvic girdle in the midventral line and carefully separate the bones and muscles to find the cloaca.
6.  Identify the urinary bladder attached to the ventral wall of the cloaca. In frogs, urine backs up into the bladder from the cloaca.

**7.** Explain the term *urogenital system.* _____

_____

**8.** The cloaca receives material from (1) _____ ,

(2) _____ , and (3) _____ .

**Figure 15.18** **Urogenital system of a male frog.**
*a.* Drawing. *b.* Dissected specimen.

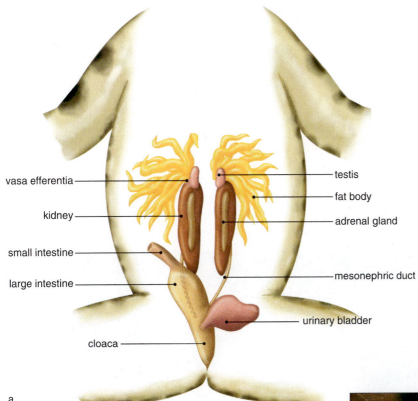

vasa efferentia
kidney
small intestine
large intestine
cloaca

testis
fat body
adrenal gland
mesonephric duct
urinary bladder

a.

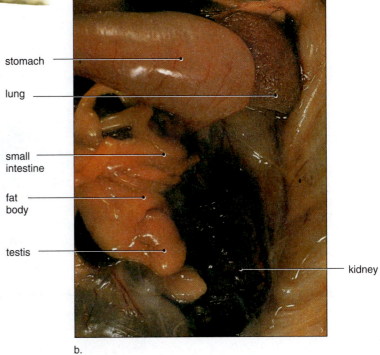

stomach
lung
small intestine
fat body
testis
kidney

b.

**Figure 15.19** Urogenital system of a female frog.
a. Drawing. b. Dissected specimen.

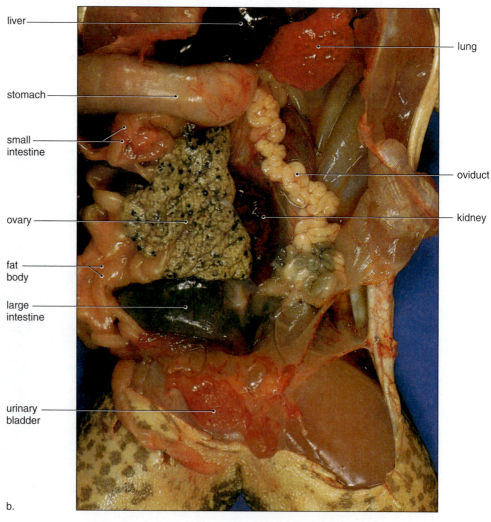

ostium

oviduct

kidney

ovary

large intestine

urinary bladder

fat body

adrenal gland

mesonephric duct

uterus

cloaca

a.

liver

stomach

small intestine

ovary

fat body

large intestine

urinary bladder

lung

oviduct

kidney

b.

# 15.4 Comparison of Invertebrates with Vertebrates

**Vertebrates** are segmented, and specialization of parts has occurred. Arthropods have an exoskeleton, while vertebrates have an endoskeleton, but they both have jointed appendages. In vertebrates, two pairs of appendages are characteristic. The vertebrate brain is more complex than that of arthropods and is enclosed by a skull. Among vertebrates, a high degree of cephalization is the rule. All organ systems are present and efficient.

Complete Table 15.7 with a "yes" or "no" to compare the invertebrates studied with the vertebrates studied in this laboratory.

| Table 15.7 | Comparison of Invertebrates with Vertebrates | | | | |
|---|---|---|---|---|---|
| | Coelom | Exoskeleton | Endoskeleton | Jointed Appendages | Location of Nerve Cord (ventral/dorsal) |
| Invertebrates | | | | | |
| | | | | | |
| | | | | | |
| | | | | | |
| | | | | | |
| | | | | | |
| | | | | | |
| | | | | | |
| Vertebrates | | | | | |
| | | | | | |
| | | | | | |
| | | | | | |
| | | | | | |
| | | | | | |
| | | | | | |

_____  1. Jointed appendages and an exoskeleton are characteristic of what group of animals?

_____  2. Crayfish belong to what group of arthropods?

_____  3. A clam belongs to what group of molluscs?

_____  4. A visceral mass, foot, and mantle are characteristic of what group of animals?

_____  5. In a clam, which structure secretes the shell?

_____  6. The arthropods are the first of the animal phyla to have what general characteristic?

_____  7. Contrast the respiratory organ of a crayfish with that of a grasshopper.

_____  8. In a frog, the glottis allows air to enter the _____.

_____  9. In a frog, the esophagus allows food to enter the _____.

_____ 10. In a frog, the cloaca receives material from the intestine, the kidneys, and the _____.

## Thought Questions

11. Compare respiratory organs in the crayfish and the grasshopper. How are these suitable to the habitat of each?

12. For each of the following characteristics, name an animal with the characteristic, and state the characteristic's advantages:

   a. Presence of a head region

   b. Jointed appendages

# 16

# Basic Mammalian Anatomy I

## Learning Objectives

**16.1 External Anatomy**
- Compare the limbs of a pig to the limbs of a human.
- Identify the sex of a fetal pig.

**16.2 Oral Cavity and Pharynx**
- Find and identify the teeth; tongue; and hard and soft palates, including the uvula.
- Identify and state a function for the glottis, nasopharynx, and esophagus.
- Name the two pathways that cross in the pharynx.

**16.3 Thoracic and Abdominal Incisions**
- List the major organs of the thoracic cavity and the major organs of the abdominal cavity.

**16.4 Neck Region**
- Find, identify, and state a function for the thymus, the larynx, and the thyroid gland.

**16.5 Thoracic Cavity**
- Identify the three compartments and the organs of the thoracic cavity.
- Find and identify the diaphragm.

**16.6 Abdominal Cavity**
- Find, identify, and state a function for the liver, stomach, spleen, gallbladder, pancreas, large intestine, and small intestine. Describe where these organs are positioned in relation to one another.

## Introduction

In this laboratory, you will dissect a fetal pig. Both pigs and humans are mammals; therefore, you will be studying mammalian anatomy. The period of pregnancy, or gestation, in pigs is approximately 17 weeks (compared to an average of 40 weeks in humans). The piglets used in class will usually be within 1 to 2 weeks of birth.

The pigs will have a slash in the right neck region, indicating the site of blood drainage. A red latex solution was injected into the **arterial system,** and a blue latex solution was injected into the **venous system** of the pigs. Therefore, when a vessel appears red, it is an artery, and when a vessel appears blue, it is a vein.

> **Caution:** Wear protective latex gloves when handling preserved animal organs.

# 16.1 External Anatomy

Mammals are characterized by the presence of mammary glands and hair. Mammals also occur in two distinct sexes, called males and females, often distinguishable by their external **genitals,** the reproductive organs.

Both pigs and humans are placental mammals, which means that development occurs within the uterus of the mother. An **umbilical cord** stretches externally between the fetal animal and the **placenta,** where carbon dioxide and organic wastes are exchanged for oxygen and organic nutrients.

Pigs and humans are tetrapods—that is, they have four limbs. Pigs walk on all four of their limbs; in fact, they walk on their toes, and their toenails have evolved into hooves. In contrast, humans walk only on the feet of their hind limbs.

## Observation: External Anatomy

### Body Regions and Limbs

1. Place your animal in a dissecting pan, and observe the following body regions: the rather large head; the short, thick neck; the cylindrical trunk with two pairs of appendages (forelimbs and hindlimbs); and the short tail (Fig. 16.1*a*). The tail is an extension of the vertebral column.
2. Examine the four limbs, and feel for the joints of the digits, wrist, elbow, shoulder, hip, knee, and ankle.
3. Determine which parts of the forelimb correspond to your upper arm, elbow, lower arm, wrist, and hand.
4. Do the same for the hind limb, comparing it to your leg.
5. The pig walks on its toenails, which would be like a ballet dancer on "tiptoe." Notice how your heel touches the ground when you walk. Where is the heel of the pig? _____

### Umbilical Cord

1. Locate the umbilical cord arising from the ventral (toward the belly) portion of the abdomen.
2. Note the cut ends of the umbilical blood vessels. If they are not easily seen, cut the umbilical cord near the end and observe this new surface.
3. What is the function of the umbilical cord? _____

_____

### Nipples and Hair

1. Locate the small **nipples,** the external openings of the **mammary glands.** The nipples are *not* an indication of sex, since both males and females possess them. How many nipples does your pig have? _____ When is it advantageous for a pig to have so many nipples? _____

_____

2. Can you find hair on your pig? _____ Where? _____

_____

## Figure 16.1 External anatomy of the fetal pig.

*a.* Body regions and limbs. *b, c.* The sexes can be distinguished by the external genitals.

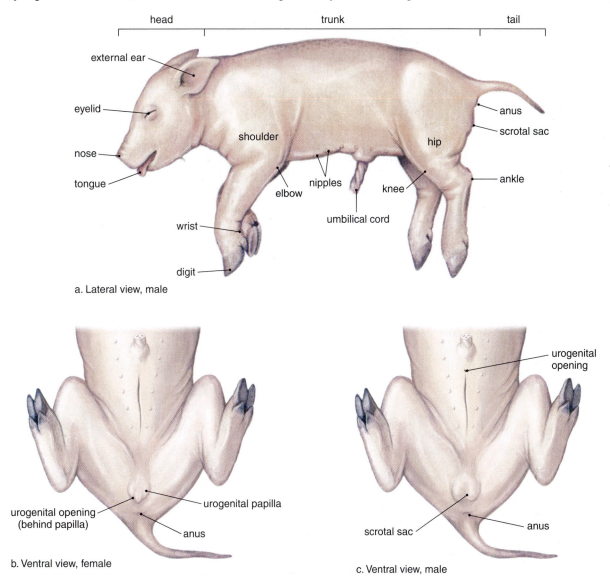

a. Lateral view, male

b. Ventral view, female

c. Ventral view, male

### Anus and External Genitals

1. Locate the **anus** under the tail. The anus is an opening for what system in the body? _____
2. In females, locate the **urogenital opening,** just anterior to the anus, and a small, fleshy **urogenital papilla** projecting from the urogenital opening (Fig. 16.1*b*).
3. In males, locate the urogenital opening just posterior to the umbilical cord. The duct leading to it runs forward from between the legs in a long, thick tube, the **penis,** which can be felt under the skin. In males, the urinary system and the genital system are always joined (Fig. 16.1*c*).
4. You are responsible for identifying pigs of both sexes. What sex is your pig? _____ Be sure to look at a pig of the opposite sex that another group of students is dissecting.

# 16.2 Oral Cavity and Pharynx

The **oral cavity** is the space in the mouth that contains the tongue and the teeth. The **pharynx** is dorsal to the oral cavity and has three openings: The **glottis** is an opening through which air passes on its way to the **trachea** (the windpipe) and lungs. The **esophagus** is a portion of the digestive tract that leads through the neck and thorax to the stomach. The **nasopharynx** leads to the nasal passages.

### Observation: Oral Cavity and Pharynx

*Oral Cavity*

1. Insert a sturdy pair of scissors into one corner of the specimen's mouth, and cut posteriorly (toward the hind end) for approximately 4 cm. Repeat on the opposite side.
2. Place your thumb on the tongue at the front of the mouth, and gently push downward on the lower jaw. This will tear some of the tissue in the angles of the jaws so that the mouth will remain partly open (Fig. 16.2).
3. Note small, underdeveloped teeth in both the upper and lower jaws. Other embryonic nonerupted teeth may also be found within the gums. The teeth are used to chew food.
4. Examine the tongue, which is partly attached to the lower jaw region but extends posteriorly and is attached to a bony structure at the back of the oral cavity (Fig. 16.2). The tongue manipulates food for swallowing.
5. Locate the hard and soft palates (Fig. 16.2). The **hard palate** is the ridged roof of the mouth that separates the oral cavity from the nasal passages. The **soft palate** is a smooth region posterior to the hard palate. An extension of the soft palate—the **uvula**—hangs down into the throat in humans. (A pig does not have a uvula.)

**Figure 16.2** **Oral cavity of the fetal pig.**
The roof of the oral cavity contains the hard and soft palates, and the tongue lies above the floor of the oral cavity.

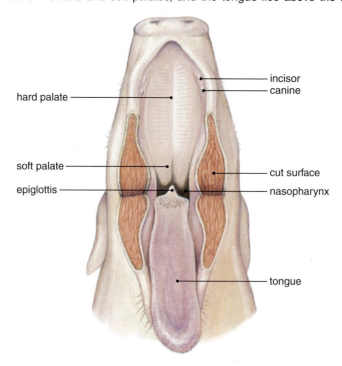

### Pharynx

1. Push down on the tongue until you open the jaws far enough to see a slightly pointed flap of tissue pointing dorsally (toward the back) (Fig. 16.2). This flap is the **epiglottis,** which covers the glottis. The glottis leads to the trachea (Fig. 16.3*a*).
2. Posterior and dorsal to the glottis, find the opening into the esophagus. Note the proximity of the glottis and the opening to the esophagus. Each time the pig—or a human—swallows, the epiglottis instantly closes to keep food and fluids from going into the lungs via the trachea.
3. Insert a blunt probe into the glottis, and note that it enters the trachea. Remove the probe, insert it into the esophagus, and note the position of the esophagus beneath the trachea.
4. Make a midline cut in the soft palate from the epiglottis to the hard palate. Then make two lateral cuts at the edge of the hard palate.
5. Posterior to the soft palate, locate the openings to the nasal passages.
6. Explain why it is correct to say that the air and food passages cross in the pharynx.

_____

_____

### Figure 16.3  Air and food passages in the fetal pig.
The air and food passages cross in the pharynx. *a.* Drawing. *b.* Dissection of specimen.

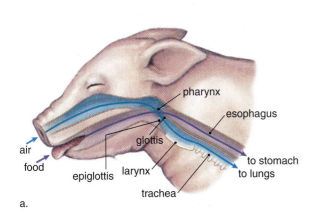

a.

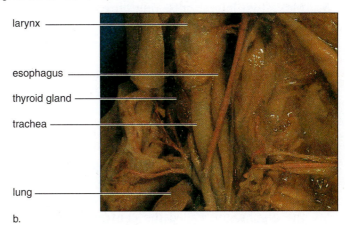

b.

# 16.3 Thoracic and Abdominal Incisions

First, prepare your pig according to the following directions, and then make thoracic and abdominal incisions so that you will be able to study the internal anatomy of your pig.

## Preparation of Pig for Dissection

1. Place the fetal pig on its back in the dissecting pan.
2. Tie a cord around one forelimb, and then bring the cord around underneath the pan to fasten back the other forelimb.
3. Spread the hindlimbs in the same way.
4. With scissors always pointing up (never down), make the following incisions to expose the thoracic and abdominal cavities. The incisions are numbered on Figure 16.4 to correspond with the following steps.

### Thoracic Incisions

1. Cut anteriorly up from the **diaphragm**, a structure that separates the thoracic cavity from the abdominal cavity, until you reach the hairs in the throat region.
2. Make two lateral cuts, one on each side of the midline incision anterior to the forelimbs, taking extra care not to damage the blood vessels around the heart.
3. Make two lateral cuts, one on each side of the midline just posterior to the forelimbs and anterior to the diaphragm, following the ends of the ribs. Pull back the flaps created by these cuts to expose the **thoracic cavity.** List the organs you find in the thoracic cavity.

_____

### Abdominal Incisions

4. With scissors pointing up, cut posteriorly from the diaphragm to the umbilical cord.
5. Make a flap containing the umbilical cord by cutting a semicircle around the cord and by cutting posteriorly to the left and right of the cord.
6. Make two cuts, one on each side of the midline incision posterior to the diaphragm. Examine the diaphragm, attached to the chest wall by radially arranged muscles. The central region of the diaphragm, called the **central tendon,** is a membranous area.
7. Make two more cuts, one on each side of the flap containing the umbilical cord and just anterior to the hind limbs. Pull back the side flaps created by these cuts to expose the **abdominal cavity.**
8. Lifting the flap with the umbilical cord requires cutting the **umbilical vein.** Before cutting the umbilical vein, tie a thread on each side of where you will cut to mark the vein for future reference.
9. Rinse out your pig as soon as you have opened the abdominal cavity. If you have a problem with excess fluid, obtain a disposable plastic pipette to suction off the liquid.
10. Anatomically, the diaphragm separates what two cavities?

_____

11. List the organs you find in the abdominal cavity.

_____

**Figure 16.4  Ventral view of the fetal pig indicating incisions.**

These incisions are to be made preparatory to dissecting the internal organs. They are numbered here in the order they should be done.

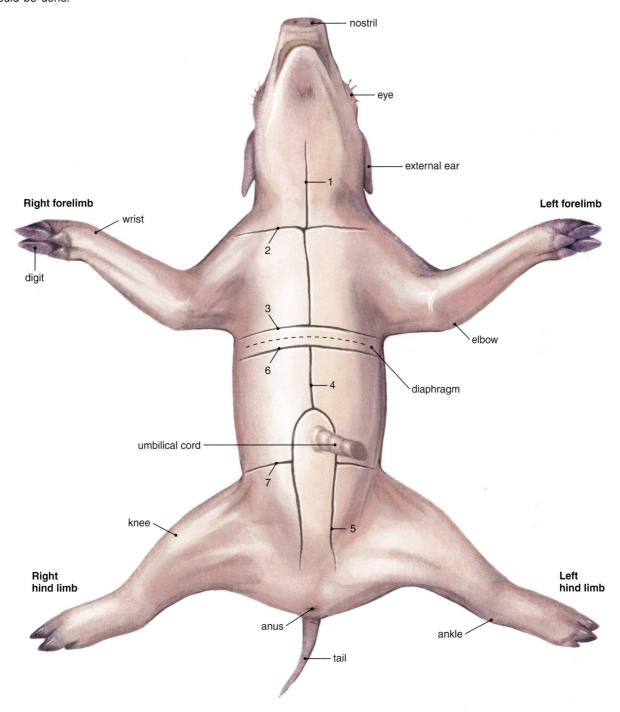

## 16.4 Neck Region

You will locate several organs in the neck region. Use Figure 16.3b as a guide, but **keep all the flaps on your pig** so you can close the thoracic and abdominal cavities at the end of the laboratory session.

The **thymus gland** is a part of the lymphatic system. Certain white blood cells called T (for thymus) lymphocytes mature in the thymus gland and help fight disease. The **larynx,** or voice box, sits atop the **trachea,** or windpipe. The **esophagus** is a portion of the digestive tract that leads to the stomach. The **thyroid gland** secretes hormones that travel in the blood and act upon other body cells. These hormones (e.g., thyroxine) regulate the rate at which metabolism occurs in cells.

### Observation: Neck Region

#### Thymus Gland

1. Move the skin apart in the neck region just below the hairs mentioned earlier. If necessary, cut the body wall laterally to make flaps. You will most likely be viewing exposed muscles.
2. *Cut through and clear away muscle* to expose the *thymus gland,* a diffuse gland that lies among the muscles. Later you will notice that the thymus flanks the thyroid and overlies the heart. The thymus is particularly large in fetal pigs, since their immune systems are still developing.

#### Larynx, Trachea, and Esophagus

1. Probe down into the deeper layers of the neck. Medially (toward the center), beneath several strips of muscle, you will find the hard-walled larynx and the trachea, parts of the respiratory passage to be examined later. Dorsal to the trachea, find the esophagus.
2. Open the mouth and insert a probe into the glottis and esophagus from the pharynx to better understand the orientation of these two organs.

#### Thyroid Gland

Locate the thyroid gland just posterior to the larynx, lying ventral to (on top of) the trachea.

## 16.5 Thoracic Cavity

As previously mentioned, the body cavity of mammals, including human beings, is divided by the diaphragm into the thoracic cavity and the abdominal cavity. The heart and lungs are in the thoracic cavity (Fig. 16.5). The **heart** is a pump for the cardiovascular system, and the **lungs** are organs of the respiratory system where gas exchange occurs.

### Observation: Thoracic Cavity

#### Heart and Lungs

1. If you have not yet done so, fold back the chest wall flaps. To do this, you will need to tear the thin membranes that divide the thoracic cavity into three compartments: the left **pleural cavity** containing the left lung, the right pleural cavity containing the right lung, and the **pericardial cavity** containing the heart.
2. Examine the lungs. Locate the four lobes of the right lung and the three lobes of the left lung. The trachea, dorsal to the heart, divides into the **bronchi,** which enter the lungs. Later, when the heart is removed, you will be able to see the trachea and bronchi.
3. Trace the path of air from the nasal passages to the lungs.

_____

_____

# 16.6 Abdominal Cavity

The abdominal wall and organs are lined by a membrane called **peritoneum,** consisting of epithelium supported by connective tissue. Double-layered sheets of peritoneum, called **mesenteries,** project from the body wall and support the organs.

The **liver,** the largest organ in the abdomen (Fig. 16.5), performs numerous vital functions, including (1) disposing of worn-out red blood cells, (2) producing bile, (3) storing glycogen, (4) maintaining the blood glucose level, and (5) producing blood proteins.

The abdominal cavity also contains organs of the digestive tract, such as the stomach, small intestine, and large intestine. The **stomach** stores food and has numerous gastric glands that secrete gastric juice, which digests protein. The **small intestine** is a part of the digestive tract that receives secretions from the pancreas and gallbladder. Besides being an area for the digestion of all components of food—carbohydrate, protein, and fat—the small intestine absorbs the products of digestion: glucose, amino acids, glycerol, and fatty acids. The **large intestine** is a part of the digestive tract that absorbs water and prepares feces for defecation at the anus.

The **gallbladder** stores and releases bile, which aids the digestion of fat. The **pancreas** is both an exocrine and an endocrine gland. As an exocrine gland, it produces and secretes pancreatic juice, which digests all the components of food in the small intestine. Both bile and pancreatic juice enter the duodenum by way of ducts. As an endocrine gland, the pancreas secretes the hormones insulin and glucagon into the bloodstream. Insulin and glucagon regulate blood glucose levels.

The **spleen** is a lymphoid organ in the lymphatic system that contains both white and red blood cells. It purifies blood and disposes of worn-out red blood cells.

## Observation: Abdominal Cavity

### Liver

1. If your particular pig is partially filled with dark, brownish material, take your animal to the sink and rinse it out. This material is clotted blood. Consult your instructor before removing any red or blue latex masses, since they may enclose organs you will need to study.
2. Locate the liver, a large brown organ. Its anterior surface is smoothly convex and fits snugly into the concavity of the diaphragm.
3. Name several functions of the liver. _____

_____

### Stomach and Spleen

1. Push aside and identify the stomach, a large sac dorsal to the liver on the left side.
2. Locate the point near the midline of the body where the **esophagus** penetrates the diaphragm and joins the stomach.
3. Find the spleen, a long, flat reddish organ attached to the stomach by mesentery.
4. The stomach is a part of what system? _____

   What is its function? _____
5. The spleen is a part of what system? _____

   What is its function? _____

# Figure 16.5 Internal anatomy of the fetal pig.

Most of the major organs are shown in this photograph. The stomach has been removed. The spleen, gallbladder, and pancreas are not visible. *Do not* remove any organs or flaps from your pig.

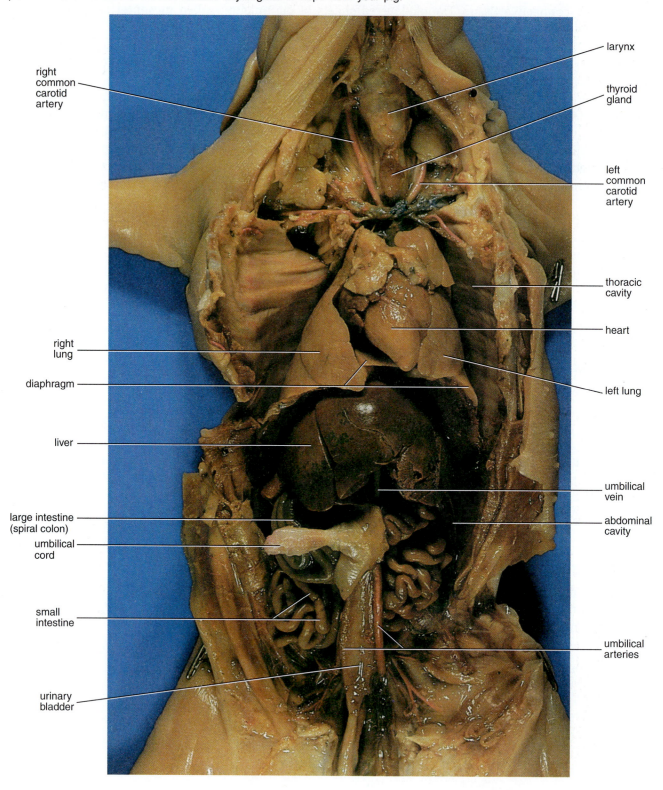

right common carotid artery

larynx

thyroid gland

left common carotid artery

thoracic cavity

heart

right lung

diaphragm

left lung

liver

umbilical vein

large intestine (spiral colon)

abdominal cavity

umbilical cord

small intestine

umbilical arteries

urinary bladder

### Small Intestine

1. Look posteriorly where the stomach makes a curve to the right and narrows to join the anterior end of the small intestine called the **duodenum.**
2. From the duodenum, the small intestine runs posteriorly for a short distance and is then thrown into an irregular mass of bends and coils held together by a common mesentery.
3. The small intestine is a part of what system? _____

   What is its function? _____

### Gallbladder and Pancreas

1. Locate the **bile duct,** which runs in the mesentery stretching between the liver and the duodenum. Find the gallbladder, embedded in the liver on the underside of the right lobe. It is a small, greenish sac.
2. Lift the stomach and locate the pancreas, the light-colored, diffuse gland lying in the mesentery between the stomach and the small intestine. The pancreas has a duct that empties into the duodenum of the small intestine.
3. What is the function of the gallbladder? _____
4. What is the function of the pancreas? _____

   _____

### Large Intestine

1. Locate the distal (far) end of the small intestine, which joins the large intestine posteriorly, in the left side of the abdominal cavity (right side in humans). At this junction, note the **cecum,** a blind pouch.
2. Follow the main portion of the large intestine, known as the **colon,** as it runs from the point of juncture with the small intestine into a tight coil (spiral colon), then out of the coil anteriorly, then posteriorly again along the midline of the dorsal wall of the abdominal cavity. In the pelvic region, the **rectum** is the last portion of the large intestine. The rectum leads to the **anus.**
3. The large intestine is a part of what system? _____
4. What is the function of the large intestine? _____
5. Trace the path of food from the mouth to the anus. _____

   _____

---

### Storage of Pigs

1. Before leaving the laboratory, place your pig in the plastic bag provided.
2. Expel excess air from the bag, and tie it shut.
3. Write your *name* and *section* on the tag provided, and attach it to the bag. Your instructor will indicate where the bags are to be stored until the next laboratory period.
4. Clean the dissecting tray and tools, and return them to their proper location.
5. Wipe off your goggles.
6. Wash your hands.

_____  1. In the fetal pig, what sex has a urogenital opening beneath the papilla just superior to the anus?

_____  2. What two characteristics do all mammals have?

_____  3. The esophagus connects the pharynx with which organ?

_____  4. What is the hard portion of the roof of the mouth called?

_____  5. What is the opening to the trachea called?

_____  6. Name the largest organ in the abdominal cavity.

_____  7. What structure separates the thoracic cavity from the abdominal cavity?

_____  8. Name the structure just dorsal to the thyroid gland.

_____  9. What structure covers the glottis?

_____  10. If a probe is placed through the glottis, it will enter what structure?

_____  11. The heart is located in what cavity?

_____  12. What organs are in the pleural cavity?

_____  13. The stomach connects to what part of the small intestine?

_____  14. Identify the gland located by lifting the stomach.

_____  15. Name a lymphoid organ in the abdominal cavity.

_____  16. Is the spleen located on the pig's right or left side?

_____  17. The pancreas belongs to what system of the body?

## Thought Questions

18. What difficulty would probably arise if a person were born without an epiglottis?

19. A large portion of the abdominal cavity is taken up by digestive organs. Which organs are these?

# 17

# Cardiovascular System

## Learning Objectives

**17.1 The Heart**
- Locate and identify the chambers of the heart and their attached blood vessels.
- Name and locate the valves of the heart.
- Trace the path of blood through the heart.
- Find and identify the coronary arteries and cardiac veins.

**17.2 Path of Blood in the Pulmonary and Systemic Circuits**
- Find and identify the pulmonary blood vessels, and trace the path of blood from the heart to and from the lungs.
- Find and identify the vessels in the thoracic cavity that allow you to trace the path of blood from the heart to and from the head and to and from the forelimbs.
- Find and identify the vessels in the abdominal cavity that allow you to trace the path of blood from the heart to and from the kidneys and to and from the hindlimbs.

**17.3 Heart Rate and Blood Pressure**
- Determine the heartbeat by taking the pulse.
- Correlate increased heart rate with exercise and increased blood pressure with increased heart rate.

## Introduction

Blood must circulate to serve the body. The heart pumps the blood, which moves away from the heart in **arteries** and **arterioles** and returns to the heart in **venules** and **veins. Capillaries** connect arterioles to venules.

In today's laboratory, you will study the anatomy of the human heart and how blood circulates through the heart. The heart is a double pump, the right side pumps the blood into the pulmonary circuit (to the lungs); and the left side pumps the blood into the systemic circuit (to the body). You will dissect the major blood vessels in the pulmonary and systemic circuits as a preliminary to learning to trace the path of blood in these circuits.

The heart is a pump because when it contracts, blood is sent out under pressure. This pressure can be felt as a pulsation, or throb, in the arteries of the wrist with each heartbeat. Taking the pulse tells you the heart rate. Exercise increases the heart rate because skeletal muscle contraction raises the carbon dioxide and $H^+$ levels of the blood. This change is detected by sensory receptors in certain arteries and, thereafter, the cardiovascular center in the brain directs an increased heart rate (Fig. 17.1).

**Figure 17.1** The human cardiovascular system.

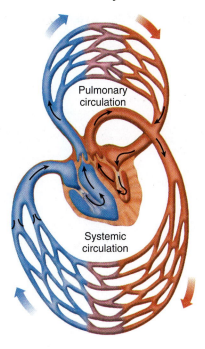

Pulmonary circulation

Systemic circulation

---

**Caution:** Wear protective latex gloves when handling preserved animal organs.

---

# 17.1 The Heart

The mammalian heart has a right and left atrium and a right and left ventricle. *To tell the left from the right side, mentally position the heart so it corresponds to your body.* There is at least one blood vessel attached to each of the chambers. The heart valves keep the blood moving forward because backward flow closes the valves. The heart nodes cause the heart to contract every .85 second. The contraction of the heart pumps the blood through the heart and out into the arteries. The right ventricle pumps blood into the pulmonary trunk, which leads to the pulmonary arteries, and the left ventricle pumps blood into the aorta, the major artery in the body. The heart is a muscle that has its own blood supply. The coronary arteries, which lie on the surface of the heart, arise from the aorta. The cardiac veins drain into the right atrium of the heart.

## Heart Model or Preserved Sheep Heart

The heart model or preserved sheep heart will be used to study the anatomy of the heart (Fig. 17.2). If examining a preserved sheep heart (Fig. 17.2b), begin by finding the ventral (front) and dorsal (back) of the heart.

### Observation: External Anatomy

1. Identify the **right atrium,** and its attached blood vessels, the superior (anterior) and inferior (posterior) **venae cavae.** The superior vena cava and the inferior vena cava return blood from the head and body, respectively, to the right atrium.
2. Identify the **right ventricle** and its attached blood vessel, the **pulmonary trunk.** The pulmonary trunk leaves the ventral side of the heart from the top of the right ventricle and then passes forward diagonally before branching into the **right** and **left pulmonary arteries.**
3. Identify the **left atrium** and its attached blood vessels, the left and right **pulmonary veins.** The pulmonary veins return blood from the lungs to the left atrium.

**Figure 17.2 External heart anatomy.**
*a.* Human heart. *b.* Sheep heart.

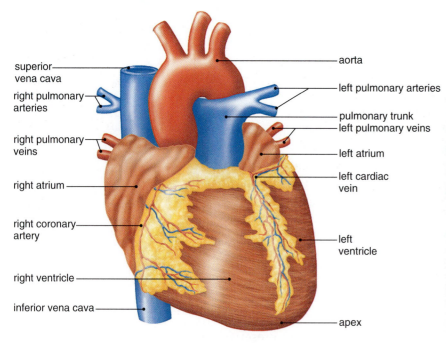

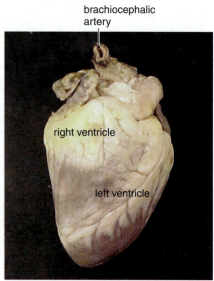

a.

b.

4. Identify the **left ventricle** and its attached blood vessel; the **aorta,** the main systemic blood vessel, which arises from the anterior end of the left ventricle, just dorsal to the origin of the pulmonary trunk. The aorta soon bends to the animal's left as the aortic arch. The aorta carries blood to the body proper.

5. Identify the **coronary arteries** and the **cardiac veins,** which service the needs of the heart wall. The coronary arteries branch off the aorta as soon as it leaves the heart and they appear on the surface of the heart. The cardiac veins, also on the surface of the heart, join and then enter the right atrium through the coronary sinus.

### Observation: Internal Anatomy

Remove the ventral half of the human heart model (Fig. 17.3a) or the ventral half of the preserved sheep heart (Fig. 17.3b).

1. Identify the four chambers of the heart in longitudinal section: **right atrium, right ventricle, left atrium,** and **left ventricle** and label in Fig. 17.3a.

2. Which ventricle is more muscular? _____

3. Why is this appropriate? _____

4. Identify the four main blood vessels attached to the heart and label these.

#### Valves of the Heart

1. Locate the four valves of the heart. Find the right atrioventricular, called the **tricuspid valve,** located between the right atrium and the right ventricle.

2. Find the left atrioventricular, called the **bicuspid valve,** located between the left atrium and the left ventricle.

3. Find the **pulmonary semilunar** valve, located in the base of the pulmonary trunk.

4. Find the **aortic semilunar** valve, located in the base of the aorta. What is the function of the

   heart valves? _____

5. Note the **chordae tendineae** ("heartstrings") that attach to atrioventricular valves and extend from the papillary muscles. The chordae tendineae prevent the atrioventricular valves from inverting into the atria when the ventricles contract.

## Figure 17.3  Internal heart anatomy.
*a.* Human heart. *b.* Sheep heart.

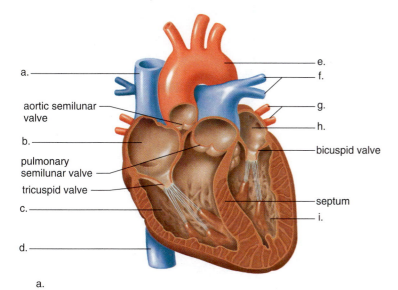

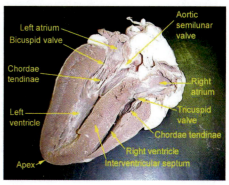

a.

b.

## Path of Blood Through the Heart

*To demonstrate that $O_2$-poor blood is kept separate from $O_2$-rich blood, trace the path of blood from the right side of the heart to the aorta by filling in the following blanks.* After birth, the blood passes through the lungs to go from the right to the left sides of the heart.

**Venae Cavae**                                    **Lungs**

_____                                  _____

_____ valve                            _____

_____                                  _____ valve

_____ valve                            _____

_____                                  _____ valve

Lungs                                              Aorta

# 17.2 Path of Blood in the Pulmonary and Systemic Circuits

In adult mammals, the heart is a double pump. The right side of the heart, consisting of two chambers (the right atrium and right ventricle), pumps blood into the **pulmonary circuit**—that is, to the lungs and back to the heart (Fig. 17.4). While the blood is in the lungs, it gives up carbon dioxide and gains oxygen. The left side of the heart, consisting of two chambers (the left atrium and left ventricle), pumps blood to the **systemic circuit**—that is, throughout the whole body except the lungs. Blood in the systemic circuit gives up oxygen and gains carbon dioxide.

Figure 17.4 shows how to trace the path of blood in both the pulmonary and systemic circuits in adult humans. It will also assist you in learning the names of the blood vessels you will be dissecting. In humans, anatomists refer to the superior and inferior venae cavae. Because the pig is a four-legged animal, these same vessels are called the anterior and posterior venae cavae.

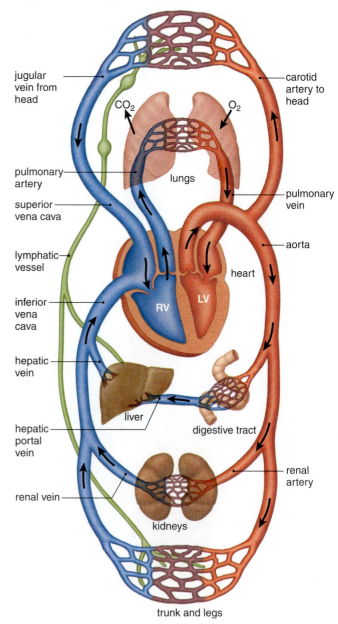

**Figure 17.4** **Diagram of the human cardiovascular system.**

In the pulmonary circuit, the pulmonary arteries take $O_2$-poor blood to the lungs, and the pulmonary veins return $O_2$-rich blood to the heart. In the systemic circuit, the aorta branches into the various arteries that go to all other parts of the body. After blood passes through arterioles, capillaries, and venules, it enters various veins and then the superior (anterior) and inferior (posterior) venae cavae, which return it to the heart.

## Pulmonary Circuit

In this section, we will use the fetal pig to examine the pulmonary arteries and veins that occur in mammals, including humans. The fetal pig will have pulmonary arteries and veins, even though they are not functional until after the pig is born. In the pulmonary circuit of adult mammals, pulmonary arteries take blood away from the heart to the lungs, and pulmonary veins take blood from the lungs to the heart. Remember that in your pig, *all arteries* have been injected with red latex, and *all veins* have been injected with blue latex.

### Observation: Pulmonary Circuit

#### Pulmonary Trunk and Pulmonary Arteries

1. Locate the **pulmonary trunk**, which arises from the ventral side of the heart (see Fig. 17.2 and Fig. 17.3). It may appear white because the thick wall prevents the color of the red latex from showing through.
2. If you have not already done so, remove the pericardial sac from the heart to reveal the blood vessels entering the heart. Veins take blood to the heart.
3. Trace the pulmonary trunk. It seems to connect directly with the aorta, the major artery. This connection is the arterial duct (ductus arteriosus), unique to the fetus. The arterial duct disappears after birth.
4. In addition to this duct, look closely and you will find the **pulmonary arteries,** which leave the pulmonary trunk and go to the lungs.

#### Pulmonary Veins

1. The pulmonary veins are hard to find. If you wish, clean away the membrane dorsal to the heart, and carefully note the vessels (pulmonary veins) that leave the lungs. Trace these to the left atrium of the heart.
2. In the adult, which blood vessels—pulmonary arteries or pulmonary veins—carry $O_2$-rich blood? _____

_____

## Systemic Circuit

In adult mammals, the **systemic circuit** serves all parts of the body except the lungs. Arteries take $O_2$-rich blood from the heart to the organs, and veins take $O_2$-poor blood from the organs to the heart. The **aorta** is the major artery, and the **venae cavae** are the major veins. It will be possible for you to identify the arteries branching from the aorta and the corresponding veins branching from the venae cavae. In the fetal pig, the venae cavae are called the *anterior* vena cava and the *posterior* vena cava because the normal position of the body is horizontal rather than vertical. Use Figure 17.5 to trace blood vessels, but *do not remove any organs. You will need them in Laboratory 19.*

**Figure 17.5** **Ventral view of fetal pig arteries and veins.**
Use this diagram to trace blood vessels, but *do not remove any organs. You will need them in Laboratory 19.* The vessels of particular interest are starred. (a. = artery; v. = vein)

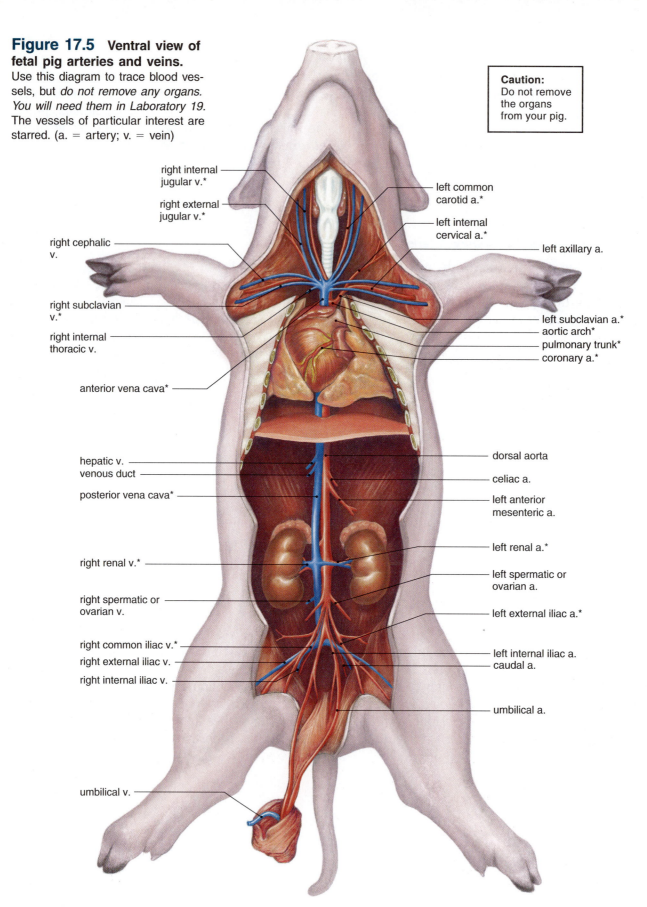

Caution:
Do not remove the organs from your pig.

right internal jugular v.*

right external jugular v.*

right cephalic v.

right subclavian v.*

right internal thoracic v.

anterior vena cava*

left common carotid a.*

left internal cervical a.*

left axillary a.

left subclavian a.*

aortic arch*

pulmonary trunk*

coronary a.*

hepatic v.
venous duct

posterior vena cava*

right renal v.*

right spermatic or ovarian v.

right common iliac v.*

right external iliac v.

right internal iliac v.

dorsal aorta

celiac a.

left anterior mesenteric a.

left renal a.*

left spermatic or ovarian a.

left external iliac a.*

left internal iliac a.

caudal a.

umbilical a.

umbilical v.

*Because the pig walks on all four limbs, the anterior vena cava in pigs is called the superior vena cava in humans, and the posterior vena cava in pigs is called the inferior vena cava in humans.

### Aorta and Venae Cavae

1. Follow the aorta as it extends through the thoracic cavity. To do this, gently move the lungs and heart to the right side of the thoracic cavity. Note the aorta; a large, whitish vessel that circles and then extends through the thoracic cavity and the diaphragm to become the **dorsal aorta.** Also note the esophagus; a smaller, flattened tube more toward the midline. The esophagus goes through the diaphragm to join with the stomach.

2. Locate the anterior (superior) vena cava coming off the top of the heart and the posterior (inferior) vena cava just to the right of the midline. The posterior vena cava also passes through the diaphragm. The posterior (inferior) vena cava is easily seen as a large, blue vessel just ventral to the dorsal aorta (see Fig. 17.8).

### Carotid Arteries and Jugular Veins

1. Find the aorta and its first branch, the **brachiocephalic artery**, which divides almost immediately into the **right subclavian** (to the pig's right shoulder) **artery** and the **carotid trunk.** The carotid trunk divides into the **right** and **left common carotid arteries** (Fig. 17.6).
2. Find the **jugular veins** alongside the carotid arteries (Fig. 17.7). Trace the carotid arteries and jugular veins as far as possible toward the head. What part of the body is serviced by the

   carotid arteries and the jugular veins? _____

   _____

3. Trace the jugular veins until they join the anterior vena cava.
4. Locate the veins that join to form the anterior vena cava (Fig. 17.7). _____

   _____

### Subclavian Arteries and Veins

1. Locate the **subclavian arteries** (Fig. 17.6) and **subclavian veins** (Fig. 17.7), which serve the upper limbs and are easily identified.
2. While the **left subclavian artery** branches from the aorta, the right subclavian artery branches from the brachiocephalic artery. In turn, the right and left subclavian arteries have branches called the right and left brachial arteries.
3. The jugular veins and the subclavian veins join to form the brachiocephalic veins.

**Figure 17.6** **Arteries of the upper body of the fetal pig.**
The vessels of particular interest are starred. (a. = artery)

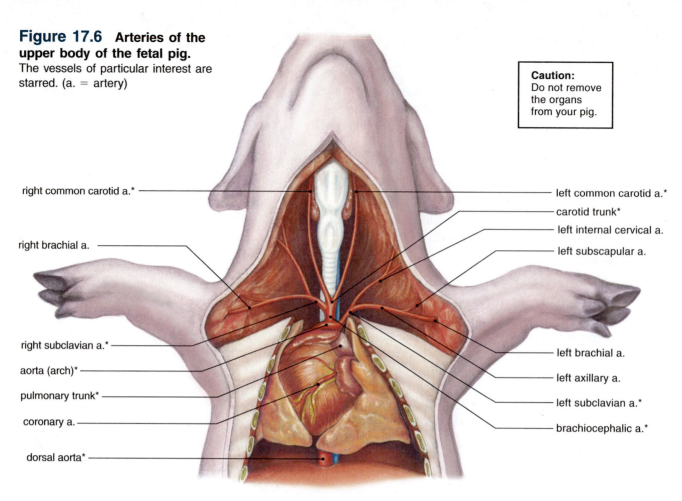

right common carotid a.*

left common carotid a.*

carotid trunk*

left internal cervical a.

left subscapular a.

right brachial a.

right subclavian a.*

aorta (arch)*

pulmonary trunk*

coronary a.

left brachial a.

left axillary a.

left subclavian a.*

brachiocephalic a.*

dorsal aorta*

**Figure 17.7** **Veins in the thoracic cavity of the fetal pig.**
The vessels of particular interest are starred. (v. = vein)

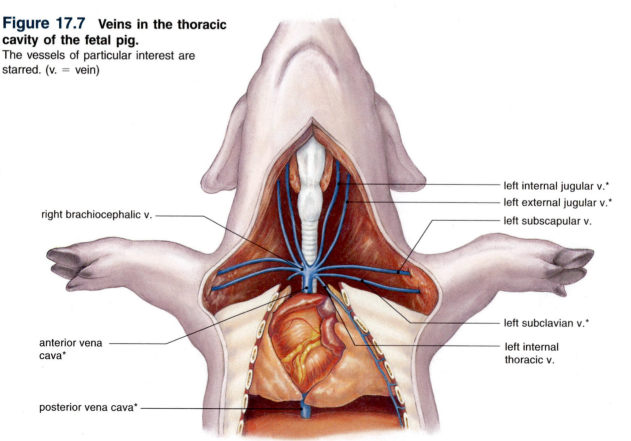

right brachiocephalic v.

left internal jugular v.*

left external jugular v.*

left subscapular v.

anterior vena cava*

left subclavian v.*

left internal thoracic v.

posterior vena cava*

## Figure 17.8 Arteries in the abdominal cavity of the fetal pig.

The vessels of particular interest are starred. (a. = artery; aa. = arteries) Use this diagram to trace blood vessels, but *do not remove any organs. You will need them in Laboratory 19.*

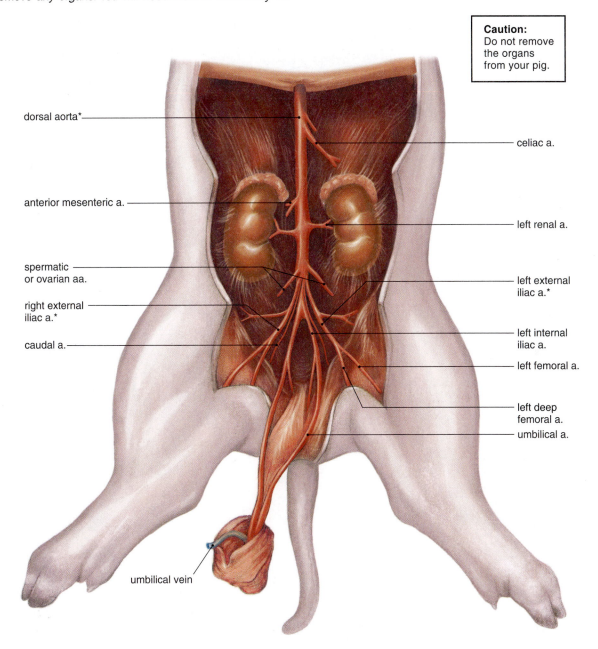

Caution:
Do not remove
the organs
from your pig.

dorsal aorta*

celiac a.

anterior mesenteric a.

left renal a.

spermatic
or ovarian aa.

left external
iliac a.*

right external
iliac a.*

left internal
iliac a.

caudal a.

left femoral a.

left deep
femoral a.

umbilical a.

umbilical vein

### Renal Arteries and Veins

1. Locate the **renal arteries** (Fig. 17.8) as they branch from the aorta, and trace these arteries as they go into the kidneys.
2. Locate the **renal veins** as they leave the kidneys (Fig. 17.9), and trace these veins as they join the posterior vena cava.

### Iliac Arteries and Veins

1. At its posterior end, the aorta branches into the paired iliac arteries (Fig. 17.8). Locate the **right** and **left external iliac arteries** at the posterior end of the aorta, and trace these arteries into the hind limbs (Fig. 17.9).
2. Find the **right** and **left common iliac veins** alongside the iliac arteries, and trace these veins as they join the posterior vena cava (Fig. 17.9).

## Figure 17.9 Veins in the abdominal cavity of the fetal pig.

The vessels of particular interest are starred. (v. = vein; vv. = veins) Use this diagram to trace blood vessels, but *do not remove any organs. You will need them in Laboratory 19.*

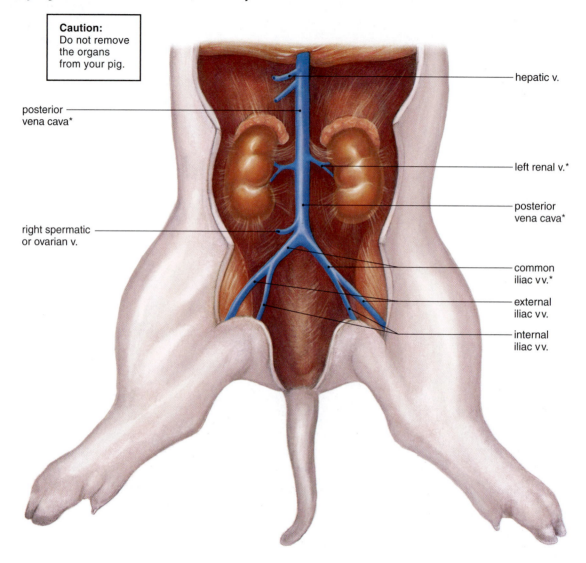

Caution:
Do not remove the organs from your pig.

posterior vena cava*

right spermatic or ovarian v.

hepatic v.

left renal v.*

posterior vena cava*

common iliac vv.*

external iliac vv.

internal iliac vv.

### Posterior Vena Cava
1. The **posterior vena cava** seems to disappear in the region of the liver. Here the posterior vena cava receives the hepatic veins coming from the liver. Scrape away some of the liver tissue to see these veins.
2. Locate the posterior vena cava as it passes through the diaphragm into the thoracic cavity and as it enters the right atrium.
3. Trace the posterior vena cava from the iliac veins to the right atrium of the heart (Fig. 17.9).

### Storage of Pigs
1. Place your pig in the plastic bag provided.
2. Expel excess air from the bag, and tie it shut.
3. Write your *name* and *section* on the tag provided, and attach it to the bag. Your instructor will indicate where the bags are to be stored until the next laboratory period.
4. Clean the dissecting tray and tools, and return them to their proper location.
5. Wipe off your goggles.
6. Wash your hands.

1. Trace the path of blood in the pulmonary circuit from the heart to the lungs, and then from the lungs to the heart.

right ventricle of heart → _____ → lungs ⟶ _____ →_____ of heart

2. Trace the path of blood in the systemic circuit from the heart to the kidneys, and then from the kidneys to the heart.

left ventricle of heart → _____→ _____ → kidneys → _____ → _____ → _____ of heart

3. Complete Table 17.1.

| Table 17.1 | Major Blood Vessels in the Systemic Circuit | |
|---|---|---|
| Body Part | Artery | Vein |
| Heart (left side) | | |
| Heart (right side) | | |
| Head | | |
| Arms | | |
| Kidney | | |
| Legs | | |

# 17.3 Heart Rate and Blood Pressure

During a heartbeat, first the atria contract and then the ventricles contract. When a chamber contracts, it is called **systole,** and when a chamber relaxes, it is called **diastole.** Blood pressure in the systemic circuit is highest just after ventricular systole S and lowest ventricular in diastole.

## Heart Rate

In the following procedure, you will take a pulse to determine the heart rate at rest and after exercise. Both of these exercises require that you work with a laboratory partner. The normal resting heart rate in a young adult is between 60 and 80 beats per minute.

### Experimental Procedure: Heart Rate

#### Heart Rate at Rest

1. To measure the pulse, position the fingers of one hand over the large artery near the outer (thumb) side of your partner's arm so that the little finger is slightly beyond the wrist. Count the pulse rate for 15 seconds, and then multiply by 4.

$$\underline{\hspace{2cm}} \times 4 = \underline{\hspace{2cm}}$$

Record your partner's pulse rate in Table 17.2.

**2.** Now switch, and your partner will determine your pulse rate.

_____ × 4 = _____

Record your pulse rate in Table 17.2.

| Table 17.2 | Heart Rate at Rest | |
|---|---|---|
| **Method** | **Partner** | **Yourself** |
| Pulse rate | | |

### Heart Rate After Exercise

1. Jump on each foot 20 times. Then, using the same method as before, determine your heartbeat after exercise, and complete Table 17.3.
2. Why is it advantageous to have an increased heart rate during exercise?

_____

| Table 17.3 | Heart Rate at Rest and After Exercise | | |
|---|---|---|---|
| **Before Exercise (from Table 17.2)** | | **After Exercise** | |
| Partner | Yourself | Partner | Yourself |
| | | | |

---

## Blood Pressure

During the following procedure, you will use a digital monitor to determine your blood pressure. The normal resting blood pressure readings for a young adult are 120/80 (systolic/diastolic), as displayed on the monitor shown in Figure 17.10.

### Figure 17.10  Measurement of blood pressure and pulse rate.

There are many different types of digital blood pressure/pulse monitors now available. The one shown here uses a cuff to be placed on the arm. Others use a cuff for the wrist.

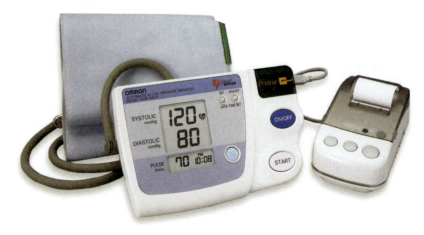

## Experimental Procedure: Blood Pressure at Rest and After Exercise

You may work with a partner or by yourself. If working with a partner, each of you will assist the other in taking blood pressure readings.

### Blood Pressure at Rest
1. Reduce your activity as much as possible.
2. Use the digital monitor to obtain blood pressure and pulse rate readings. Record your data in Table 17.4. Are the pulse readings consistent with the blood pressure readings? Offer an explanation in Table 17.4.

| Table 17.4 | Blood Pressure At Rest | | |
|---|---|---|---|
| | Blood Pressure | Pulse Rate | Explanation |
| Partner | | | |
| Yourself | | | |

### Blood Pressure After Exercise
1. Run in place for one minute.
2. Use the digital monitor to obtain blood pressure and pulse rate readings. Record your data in Table 17.5. Are the pulse readings consistent with the blood pressure readings? Offer an explanation in Table 17.5.

| Table 17.5 | Blood Pressure After Exercise | | |
|---|---|---|---|
| | Blood Pressure | Pulse Rate | Explanation |
| Partner | | | |
| Yourself | | | |

### Heart Rate and Blood Pressure
1. Blood pressure is highest just after ventricular systole, and it is lowest during ventricular diastole. Why? _____
2. Explain why you would expect a person to have a lower blood pressure reading at rest than after exercise. (In your answer, relate pulse rate to blood pressure.) _____

_____  1.  Which chamber of the heart receives venous blood from the systemic circuit?

_____  2.  Name the vessel that conducts blood from the left ventricle.

_____  3.  The pulmonary artery leaves which chamber?

_____  4.  Identify the artery that nourishes the heart tissue.

_____  5.  Which heart chamber pumps blood throughout the body?

_____  6.  Does the pulmonary artery in adults carry $O_2$-rich or $O_2$-poor blood?

_____  7.  The coronary arteries and cardiac veins serve what organ?

_____  8.  Name the blood vessel that conducts blood to the head.

_____  9.  Name the artery that serves the kidney.

_____  10.  Name the large artery that runs dorsally along the wall of the abdominal cavity.

_____  11.  Name the arteries that take blood from the aorta to the hindlimbs.

_____  12.  What part of the human body is served by the subclavian vessels?

_____  13.  Identify the large abdominal vein that runs alongside the aorta and enters the right atrium.

_____  14.  What part of the body is not served by the systemic circuit?

_____  15.  How many times a minute does the heart normally beat in a young adult?

_____  16.  What is the normal resting blood pressure of a young adult?

## Thought Questions

17.  Do arteries always carry $O_2$-rich blood? Explain.

18.  Under what conditions in everyday life would you expect the heart rate and the blood pressure to increase? When might this be an advantage? A disadvantage?

# 18

# Chemical Aspects of Digestion

## Learning Objectives

**18.1 Protein Digestion by Pepsin**
- Describe the effect of pH on the digestive action of pepsin.
- Describe the effect of environmental temperature on the digestive action of pepsin.
- Cite the test for protein digestion.

**18.2 Fat Digestion by Pancreatic Lipase**
- Describe the role of emulsification prior to the action of lipase.
- Explain the function of a control in an experiment.
- Cite the test for lipase digestion.

**18.3 Starch Digestion by Salivary Amylase**
- Describe the digestive action of amylase.
- Describe the effect of length of time on the digestive action of amylase.
- Describe the effect of boiling on the digestive action of amylase.
- Cite the test for starch digestion.

## Introduction

**Digestion** is the process by which food is broken down by **hydrolytic reactions.** That is, water is added to a large molecule, which splits into smaller soluble molecules that can be absorbed into the bloodstream. For example, water is added until proteins are broken down into amino acids, starch is broken down into glucose, and fats are broken down into glycerol and fatty acids.

Enzymes are necessary for digestion, just as they are required for other chemical reactions in the body. The experimental procedures in this laboratory will demonstrate that **pepsin** speeds the hydrolysis of protein, that **pancreatic lipase** acts on fat, and that **salivary amylase** hydrolyzes starch. Enzymes are very specific and usually participate in only one type of reaction. According to the lock-and-key theory of enzymatic action, this is because an enzyme fits its substrate just as a key fits a lock. Enzymes must maintain an appropriate shape or configuration to take part in a reaction. Enzymes have an optimum pH that allows them to maintain their usual three-dimensional shape. Enzymatic reactions speed up with increased temperature but are destroyed by excessive heat or boiling.

The Experimental Procedures in this laboratory each contain a **control,** a sample that goes through all the steps of the experiment except the one being tested (the **experimental variable**). If all of the chemicals (materials) are performing as expected, the results of this control should be negative. In experiments such as the ones in this laboratory, however, many factors, such as the purity of the chemical reagents, the age of the reagent, and the conditions of storage, can affect the outcome. In addition, organic compounds can deteriorate upon storage, particularly if they are stored under the wrong conditions.

# 18.1 Protein Digestion by Pepsin

Certain foods, such as meat and egg whites, are rich in protein. Egg whites contain albumin, the protein used in this exercise. Protein is digested by **pepsin** in the stomach, a process described by the following reaction:

$$\text{protein} + \text{water} \xrightarrow{\text{pepsin (enzyme)}} \text{peptides}$$

The stomach has a very low pH. Does this indicate that pepsin works effectively in an acidic or a basic environment? _____

## Test for Protein Digestion

Biuret reagent is used to test for protein digestion. If digestion has not occurred, biuret reagent turns purple, indicating that protein is present. If digestion has occurred, biuret reagent turns pinkish-purple, indicating that peptides are present.

> **Caution:** Biuret reagent contains a strong solution of sodium or potassium hydroxide. Exercise care in using this reagent, and follow your instructor's directions for disposing of these tubes. If any biuret reagent should spill on your skin, rinse immediately with water.

### Experimental Procedure: Protein Digestion

With a wax pencil, number four test tubes, and mark at the 2 cm, 4 cm, 6 cm, and 8 cm levels. Fill all tubes to the 2 cm mark with *albumin solution.*

Tube 1   Fill to the 4 cm mark with *pepsin solution* and to the 6 cm mark with *0.2% HCl.* Swirl to mix, and incubate at 37°C. After 1½ hours, fill to the 8 cm mark with *biuret reagent.* Record your results in Table 18.1.

Tube 2   Fill to the 4 cm mark with *pepsin solution* and to the 6 cm mark with *0.2% HCl.* Swirl to mix, and keep at room temperature. After 1½ hours, fill to the 8 cm mark with *biuret reagent.* Record your results in Table 18.1.

Tube 3   Fill to the 4 cm mark with *pepsin solution* and to the 6 cm mark with *water.* Swirl to mix, and incubate at 37°C. After 1½ hours, fill to the 8 cm mark with *biuret reagent.* Record your results in Table 18.1.

Tube 4   Fill to the 6 cm mark with *water.* Swirl to mix, and incubate at 37°C. After 1½ hours, fill to the 8 cm mark with *biuret reagent.* Record your results in Table 18.1.

**Figure 18.1  Digestion of protein.**
Pepsin, produced by the gastric glands of the stomach, helps digest protein.

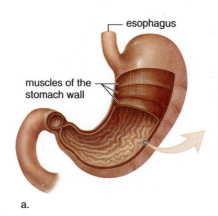

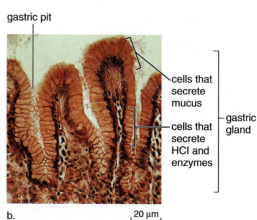

## Table 18.1 — Protein Digestion by Pepsin

| Tube | Contents | Temperature | Results of Test | Explanation |
|------|----------|-------------|-----------------|-------------|
| 1 | Albumin<br>Pepsin<br>HCl<br>Biuret reagent | | | |
| 2 | Albumin<br>Pepsin<br>HCl<br>Biuret reagent | | | |
| 3 | Albumin<br>Pepsin<br>Water<br>Biuret reagent | | | |
| 4 | Albumin<br>Water<br>Biuret reagent | | | |

## Conclusions

- Explain your results in Table 18.1 by giving a reason why digestion did or did not occur.
- Which tube was the control? _____
  Explain. _____
- If this control tube had given a positive result for protein digestion, what could you conclude about this experiment? _____

## Requirements for Digestion

Explain in Table 18.2 how each of the requirements studied in this experimental procedure influence effective digestion. Complete the table for the last two requirements after they have been studied.

## Table 18.2 — Requirements for Digestion

| Requirement | Explanation |
|-------------|-------------|
| Specific enzyme | |
| Specific substrate | |
| Warm temperature | |
| Specific pH | |
| Time | |
| Fat emulsifier | |

# 18.2 Fat Digestion by Pancreatic Lipase

Lipids include fats (e.g., butterfat) and oils (e.g., sunflower, corn, olive, and canola). Lipids are digested by **pancreatic lipase** in the small intestine, a process described by the following two reactions:

(1)
$$\text{fat} \xrightarrow{\text{bile (emulsifier)}} \text{fat droplets}$$

(2)
$$\text{fat droplets} + \text{water} \xrightarrow{\text{lipase (enzyme)}} \text{glycerol} + \text{fatty acids}$$

The first reaction is not enzymatic. It is an emulsification reaction in which fat is physically dispersed by the emulsifier (bile) into small droplets. The small droplets provide a greater surface area for enzyme action. Lipids are hydrophobic and therefore insoluble, so they are hydrolyzed slowly unless an emulsifier is used.

Given the second reaction, would the pH of the solution be lower before or after the reaction? (*Hint:* Remember that an acid decreases pH and a base increases pH.)

## Test for Fat Digestion

In the test for fat digestion, you will be using a pH indicator, which changes color as the solution in the test tube goes from basic conditions to acidic conditions. Phenol red is a pH indicator that is red in basic solutions and yellow in acidic solutions. Bile salts will be used as an emulsifier.

### Experimental Procedure: Fat Digestion

With a wax pencil, number three clean test tubes, and mark at the 1 cm, 3 cm, and 5 cm levels. Fill all the tubes to the 1 cm mark with *vegetable oil*, and to the 3 cm mark with *phenol red*.

Tube 1    Fill to the 5 cm mark with *pancreatin solution* (pancreatic lipase). Add a pinch of *bile salts*. Cover with parafilm. Invert gently to mix, and record the initial color in Table 18.3. Incubate at 37°C, and check every 20 minutes. Record the length of time for any color change.

Tube 2    Fill to the 5 cm mark with *pancreatin solution*. Invert gently to mix, and record the initial color in Table 18.3. Incubate at 37°C, and check every 20 minutes. Record the length of time for any color change.

Tube 3    Fill to the 5 cm mark with *water*. Invert gently to mix, and record the initial color in Table 18.3. Incubate at 37°C, and check every 20 minutes. Record the length of time for any color change.

| Table 18.3 | Fat Digestion by Pancreatic Lipase | | | | |
|---|---|---|---|---|---|
| Tube | Contents | Time | Color Change | | Explanation |
| | | | Initial | Final | |
| 1 | Vegetable oil Phenol red Pancreatin Bile salts | | | | |
| 2 | Vegetable oil Phenol red Pancreatin | | | | |
| 3 | Vegetable oil Phenol red Water | | | | |

## Conclusions

- Explain your results in Table 18.3 by giving a reason why digestion did or did not occur.
- What role did bile play in this experiment? _____

  _____

- What role did phenol red play in this experiment? _____
- Which test tube in this experiment could be considered a control? _____

  _____

**Figure 18.2** **Emulsification and digestion of fat.**
Bile from the liver (stored in gallbladder) enters small intestine, where lipase in pancreatic juice from the pancreas digests fat.

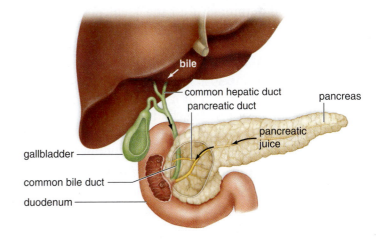

bile
common hepatic duct
pancreatic duct
pancreas
pancreatic juice
gallbladder
common bile duct
duodenum

# 18.3 Starch Digestion by Salivary Amylase

Starch is present in bakery products and in potatoes, rice, and corn. Starch is digested by **salivary amylase** in the mouth, a process described by the following reaction:

$$\text{starch} + \text{water} \xrightarrow{\text{amylase (enzyme)}} \text{maltose}$$

1. Why is this reaction called a *hydrolytic reaction*? _____
   _____

2. If digestion *does not* occur, which will be present—starch or maltose? _____

3. If digestion *does* occur, which will be present—starch or maltose? _____

## Tests for Starch Digestion

You will be using two tests for starch digestion:

1. If digestion has not taken place, the iodine test for starch will be positive. If starch is present, a blue-black color immediately appears after a few drops of iodine are added to the test tube.

2. If digestion has taken place, a test for sugar (maltose) will be positive, with red showing the highest concentration of maltose and green showing the lowest (See Table 3.4, page 36). To test for sugar, add an equal amount of Benedict's reagent to the test tube. Place the tube in a boiling water bath for 2–5 minutes, and note any color changes. A color change of blue $\longrightarrow$ green $\longrightarrow$ yellow $\longrightarrow$ orange $\longrightarrow$ red indicates the presence of maltose. Boiling the test tube is necessary for the Benedict's reagent to react. To which category of organic compounds (lipid, carbohydrate, or protein) do enzymes such as amylase belong? _____

   What happens when enzymes are boiled? _____

### Experimental Procedure: Starch Digestion

#### Preparation

1. With a wax pencil, number eight clean test tubes above the level of the boiling water bath, and mark at the 1 cm and 2 cm levels.

2. Fill tubes 1 through 6 to the 1 cm mark with *alpha-amylase solution*. Fill tubes 7 and 8 to the 1 cm mark with *water*.

3. Shake the *starch suspension* well each time before dispensing.

4. Fill tubes 3 and 4 to the 2 cm mark with *starch suspension*, and allow them to stand at room temperature for 30 minutes.

5. Place tubes 5 and 6 in a boiling water bath for 10 minutes. After boiling, fill to the 2 cm mark with *starch suspension*, and allow the tubes to stand for 20 minutes.

6. Fill tubes 7 and 8 to the 1 cm mark with *water* and to the 2 cm mark with *starch suspension*. Allow the tubes to stand for 30 minutes.

   Tube 1  Fill to the 2 cm mark with *starch suspension*, and test for starch *immediately*, using the iodine test described previously. Record your results in Table 18.4.

   Tube 2  Fill to the 2 cm mark with *starch suspension*, and test for sugar *immediately*, using *Benedict's reagent* described earlier, which requires boiling. Record your results in Table 18.4.

Why do you expect tube 1 to have a positive test for starch and tube 2 to have a negative test for sugar? _____

Record your explanation in Table 18.4.

Tubes 3, 5, and 7　　After 30 minutes, test for starch using the iodine test. Record your results in Table 18.4.

Tubes 4, 6, and 8　　After 30 minutes, test for sugar using the Benedict's test. Record your results in Table 18.4.

Why do you expect tube 3 to have a negative test for starch and tube 4 to have a positive test for sugar? _____

Why do you expect tube 5 to have a positive test for starch and tube 6 to have a negative test for sugar? _____

Why do you expect tube 7 to have a positive test for starch and tube 8 to have a negative test for sugar? _____

Record your explanations for your results in Table 18.4.

| Table 18.4 | Starch Digestion by Amylase | | | | |
|---|---|---|---|---|---|
| Tube | Contents | Time | Type of Test | Results | Explanation |
| 1 | Alpha-amylase Starch | | | | |
| 2 | Alpha-amylase Starch | | | | |
| 3 | Alpha-amylase Starch | | | | |
| 4 | Alpha-amylase Starch | | | | |
| 5 | Alpha-amylase, boiled Starch | | | | |
| 6 | Alpha-amylase, boiled Starch | | | | |
| 7 | Water Starch | | | | |
| 8 | Water Starch | | | | |

## Conclusions

- This experiment demonstrated that, for an enzymatic reaction to occur, an active _____ must be present and _____ must pass to allow the reaction to occur.

- Which test tubes served as a control in this experiment? _____
  Explain. _____

_____  1. When iodine (IKI) solution turns blue-black, what substance is present?

_____  2. What color is Benedict's reagent originally?

_____  3. What happens to an enzyme when it is boiled?

_____  4. Saliva contains what enzyme?

_____  5. As oil is digested, why does the first tube turn from red to yellow?

_____  6. What temperature promotes enzymatic action?

_____  7. What do you call a sample that goes through all the steps of an experiment but lacks the factor being tested?

_____  8. What role do bile salts play in the digestion of fat?

_____  9. What color does biuret reagent turn when peptides are present?

_____ 10. Is the optimal pH for pepsin acidic or basic?

_____ 11. Why would you predict that pepsin would not digest starch?

_____ 12. In addition to pepsin and water, what is needed to digest protein?

## Thought Questions

13. Which of the following two combinations is most likely to result in digestion?

    **a.** Pepsin, protein, water, body temperature

    **b.** Pepsin, protein, hydrochloric acid (HCl), body temperature

    Explain.

14. Which of the following two combinations is most likely to result in digestion?

    **a.** Amylase, starch, water, body temperature, testing immediately

    **b.** Amylase, starch, water, body temperature, waiting 30 minutes

    Explain.

# 19

# Basic Mammalian Anatomy II

## Learning Objectives

**19.1 Urinary System**
- Locate and identify the organs of the urinary system.
- State a function for the organs of the urinary system.

**19.2 Male Reproductive System**
- Locate and identify the organs of the male reproductive system.
- State a function for the organs of the male reproductive system.
- Compare the pig reproductive system to that of the human male.

**19.3 Female Reproductive System**
- Locate and identify the organs of the female reproductive system.
- State a function for the organs of the female reproductive system.
- Compare the pig reproductive system to that of the human female.

**19.4 Anatomy of Testis and Ovary**
- Identify a cross-section slide of the testis and seminiferous tubules, including sperm.
- Identify a slide of the ovary and the follicles, including an oocyte.

**19.5 Review of the Respiratory, Digestive, and Cardiovascular Systems**
- Identify and locate the individual organs of the respiratory, digestive, and cardiovascular systems. State a function for each organ.
- Identify and locate the hepatic portal system. State a function for this system.

## Introduction

The **urinary system** and the **reproductive system** are so closely associated in mammals that they are often considered together as the **urogenital system.** They are particularly associated in males, where certain structures function in both systems. In this laboratory, we will focus first on dissecting the urinary and reproductive systems in the fetal pig. We will then compare the anatomy of the reproductive systems in pigs to those in humans.

In mammalian reproductive systems, the testes (sing., testis) are the male gonads, and the ovaries (sing., ovary) are the female gonads. The testes produce sperm, and the ovaries produce eggs. Examining prepared slides in this laboratory will allow you to observe the location of spermatogenesis in the testis and oogenesis in the ovary.

Finally, we will examine parts of the respiratory, digestive, and cardiovascular systems in the fetal pig. You will view the organs of the respiratory system in some detail; remove and examine the heart, stomach, and intestine; and view the exposed hepatic portal system.

---

**Caution:** Wear protective latex gloves when handling preserved animal organs.

---

# Figure 19.1 Urinary system of the fetal pig.

In (*a*) females and (*b*) males, urine is made by the kidneys, transported to the bladder by the ureters, stored in the bladder, and then excreted from the body through the urethra.

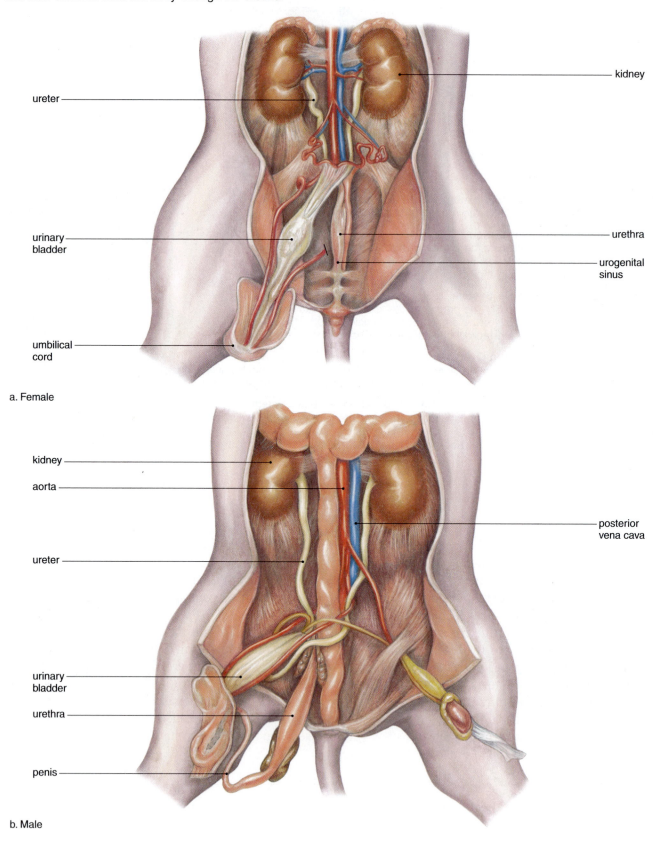

a. Female

b. Male

# 19.1 Urinary System

The urinary system consists of the **kidneys,** which produce urine; the **ureters,** which transport urine to the **urinary bladder** where urine is stored; and the **urethra,** which transports urine to the outside. In males, the urethra also transports sperm during ejaculation.

During the upcoming dissection, compare the urinary system structures of both sexes of fetal pigs. Later in this laboratory period, exchange specimens with a neighboring team for a more thorough inspection.

## Observation: Urinary System in Pigs

1. The large, paired kidneys (Fig. 19.1) are reddish organs covered by **peritoneum,** a membrane that lines the abdominal cavity. Clean the peritoneum away from one of the kidneys, and study it more closely.

2. Using a razor blade or scalpel, section one of the kidneys in place, cutting it lengthwise (Fig. 19.2). At the center of the medial portion of the kidney is an irregular, cavitylike reservoir, the **renal pelvis.** The outermost portion of the kidney (the **renal cortex**) shows many small striations perpendicular to the outer surface. This region and the more even-textured **renal medulla** region inward from it are composed of **nephrons** (excretory tubules).

3. Locate the **ureters,** which leave the kidneys and run posteriorly under the peritoneum.

4. Clean the peritoneum away, and follow a ureter to the **urinary bladder,** which normally lies in the posterior ventral portion of the abdominal cavity. The urinary bladder is on the inner surface of the flap of tissue to which the umbilical cord was attached.

5. The **urethra,** which arises from the bladder posteriorly, runs parallel to the rectum. Follow the urethra until it passes from view into the ring formed by the pelvic girdle.

6. Trace the path of urine. _____

_____

## Figure 19.2  Anatomy of the kidney.

A kidney has a renal cortex, renal medulla, renal pelvis, and microscopic tubules called nephrons. The inset shows an enlargement of a nephron.

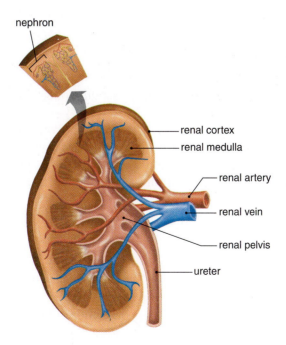

nephron

renal cortex
renal medulla
renal artery
renal vein
renal pelvis
ureter

# 19.2 Male Reproductive System

The **male reproductive system** consists of the **testes** (sing., testis), which produce sperm, and the **epididymis** (pl., epididymides), which stores sperm before they enter the **vas deferens** (pl., vasa deferentia). Just prior to ejaculation, sperm leave the vas deferens and enter the **urethra** located in the penis. The **penis** is the male organ of copulation. **Seminal vesicles,** the **prostate gland,** and the **bulbourethral glands** add fluid to semen after sperm reach the urethra. Table 19.1 summarizes the male reproductive organs.

| Table 19.1 | Male Reproductive Organs and Functions |
|---|---|
| **Organ** | **Function** |
| Testes | Produce sperm and sex hormones |
| Epididymis | Stores sperm as they mature |
| Vas deferens | Conducts and stores sperm |
| Seminal vesicle | Contributes secretions to semen |
| Prostate gland | Contributes fluid to semen |
| Urethra | Conducts sperm |
| Bulbourethral glands | Contribute mucoid fluid to semen |
| Penis | Organ of copulation |

The testes begin their development in the abdominal cavity, just anterior and dorsal to the kidneys. Before birth, however, they gradually descend into paired scrotal sacs within the **scrotum.** Each scrotal sac is connected to the body cavity by an **inguinal canal,** the opening of which can be found in your pig. The passage of the testes from the body cavity into the scrotal sacs is called the descent of the testes and also occurs in human males. The testes in most of the male fetal pigs being dissected will probably be partially or fully descended.

## Observation: Male Reproductive System in Pigs

### Inguinal Canal, Testis, Epididymis, and Vas Deferens

1. Locate the opening of the left inguinal canal, which leads to the left scrotal sac (Fig. 19.3).
2. Expose the canal and sac by making an incision through the skin and muscle layers from a point over this opening back to the left scrotal sac.
3. Open the sac, and find the testis. Note the much-coiled tubule—the epididymis—that lies alongside the testis. This is continuous with the vas deferens, which passes back toward the abdominal cavity.
4. Trace a vas deferens as it loops over an umbilical artery and ureter and unites with the urethra dorsally at the posterior end of the urinary bladder.

### Penis, Urethra, and Accessory Glands

1. Cut through the ventral skin surface just posterior to the umbilical cord. This will expose the rather undeveloped penis, which extends from this point posteriorly toward the anus. The central duct of the penis is the urethra.
2. Lay the penis to one side, and then cut down through the ventral midline, laying the legs wide apart in the process (Fig. 19.4). The cut will pass between muscles and through pelvic cartilage (bone has not developed yet). Do not cut any of the ducts or tracts in the region.
3. You will now see the urethra passing ventrally above the rectum. It is associated with certain accessory glands:
   a. Bulbourethral glands, about 1 cm in diameter, lie laterally and well back toward the anal opening.
   b. The prostate gland, about 4 mm across and 3 mm thick, is located on the dorsal surface of the urethra, just posterior to the juncture of the bladder with the urethra. It is often difficult to locate and is not shown in Figures 19.3 and 19.4.
   c. Small, paired seminal vesicles may be seen on either side of the prostate gland.

## Figure 19.3 Male reproductive system of the fetal pig.

In males, the urinary system and the reproductive system are joined. The vasa deferentia (sing., vas deferens) enter the urethra, which also carries urine.

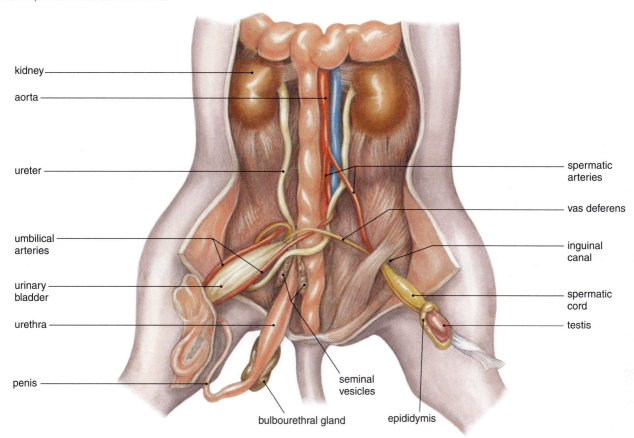

## Figure 19.4 Photograph of the male reproductive system of the fetal pig.

Compare the diagram in Figure 19.3 to this photograph to help identify the structures of the male urogenital system.

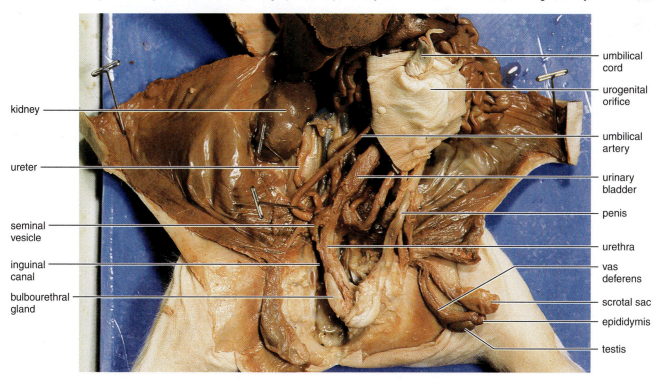

4. Trace the urethra as it leaves the bladder. It proceeds posteriorly, but when it nears the posterior end of the body, it turns rather abruptly anterioventrally and runs forward just under the skin of the midventral body wall, where you have just dissected it. This latter portion of the urethra is, then, within the penis.

5. Now you should also be able to see the entrance of the vasa deferentia into the urethra. If necessary, dissect these structures free from surrounding tissue, and expose the point of entrance of these ducts into the urethra near the location of the prostate gland. In males, the urethra transports sperm, as well as urinary wastes from the bladder.

6. Trace the path of sperm in the male. _____

## Comparison of Male Fetal Pig and Human Male

Use Figure 19.5 to help you compare the male pig reproductive system to the human male reproductive system. Complete Table 19.2, which compares the location of the penis in these two mammals.

| Table 19.2 | Location of Penis in Male Fetal Pig and Human Male | |
| --- | --- | --- |
| | **Fetal Pig** | **Human** |
| Penis | | |

## Figure 19.5 Human male urogenital system.

In the fetal pig, but not in the human male, the penis lies beneath the skin and exits at the urogenital opening.

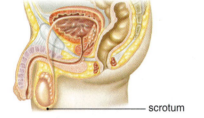

scrotum

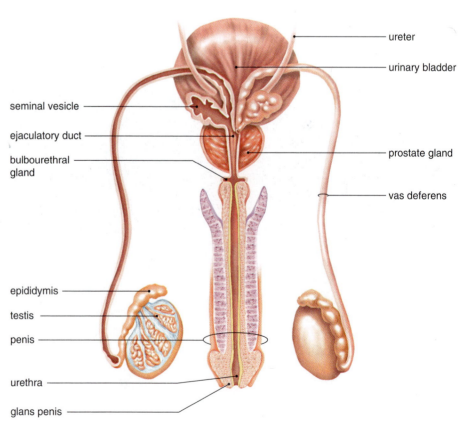

ureter

urinary bladder

seminal vesicle

ejaculatory duct

bulbourethral gland

prostate gland

vas deferens

epididymis

testis

penis

urethra

glans penis

# 19.3 Female Reproductive System

The **female reproductive system** (Table 19.3) consists of the **ovaries,** which produce eggs, and the **oviducts,** which transport eggs to the **uterus,** where development occurs. In the fetal pig, the uterus does not form a single organ, as in humans, but is partially divided into external structures called **uterine horns,** which connect with the oviduct. The **vagina** is the birth canal and the female organ of copulation.

| Table 19.3 | Female Reproductive Organs and Functions |
|---|---|
| **Organ** | **Function** |
| Ovary | Produces egg and sex hormones |
| Oviduct (fallopian tube) | Conducts egg toward uterus |
| Uterus | Houses developing fetus |
| Vagina | Receives penis during copulation and serves as birth canal |

## Observation: Female Reproductive System in Pigs

### Ovaries and Oviducts

1. Locate the paired ovaries, small bodies suspended from the peritoneal wall in mesenteries, posterior to the kidneys (Figs. 19.6 and 19.7).
2. Closely examine one ovary. Note the small, short, coiled oviduct, sometimes called the fallopian tube. The oviduct does not attach directly to the ovary but ends in a funnel-shaped structure with fingerlike processes (fimbriae) that partially enclose the ovary.

### Uterine Horns

1. Locate the **uterine horns.** (Do not confuse the uterine horns with the oviducts; the latter are much smaller and are found very close to the ovaries.)
2. Find the median body of the uterus located at the joined posterior ends of the uterine horns.

### Vagina

1. Separate the hind limbs of your specimen, and cut down along the midventral line. The cut will pass through muscle and the cartilaginous pelvic girdle. With your fingers, spread the cut edges apart, and use blunt dissecting instruments to separate connective tissue.
2. Note three ducts passing from the body cavity to the animal's posterior surface. One of these is the urethra, which leaves the bladder and passes into the **urogenital sinus.** The urethra is a part of the urinary system. The most dorsal of the three ducts is the **rectum,** which passes to its own opening, the **anus.** The rectum and anus are part of the digestive system, not the reproductive system.
3. Find the vagina, located dorsally to the urethra. The vagina is the birth canal and is also the organ of copulation. Anteriorly, it connects to the uterus, and posteriorly it enters the urogenital sinus. This sinus is absent in adult humans and several other female mammals.

## Figure 19.6 Female reproductive system of the fetal pig.

In the adult female, the urinary system and the reproductive system are separate. In the fetus, the vagina joins the urethra just before the urogenital sinus.

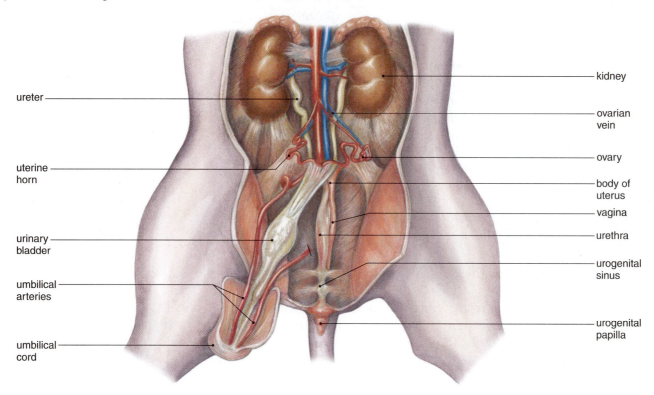

kidney

ureter

ovarian vein

ovary

uterine horn

body of uterus

vagina

urinary bladder

urethra

urogenital sinus

umbilical arteries

umbilical cord

urogenital papilla

## Figure 19.7 Photograph of the female reproductive system of the fetal pig.

Compare the diagram in Figure 19.6 to this photograph to help identify the structures of the female urinary and reproductive systems.

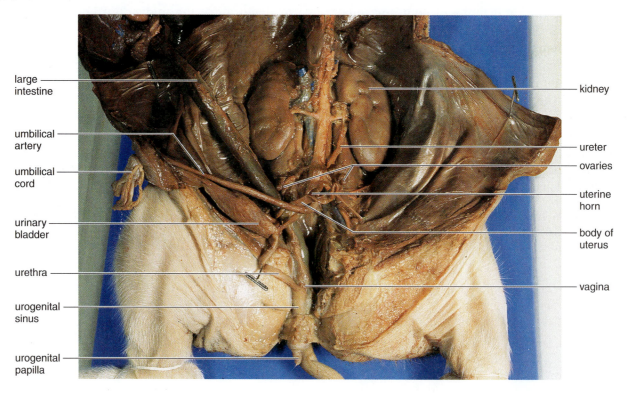

large intestine

kidney

umbilical artery

ureter

umbilical cord

ovaries

uterine horn

urinary bladder

body of uterus

urethra

vagina

urogenital sinus

urogenital papilla

## Comparison of Female Fetal Pig with Human Female

Use Figure 19.8 to compare the female pig reproductive system to the human female reproductive system. Complete Table 19.4, which compares the appearance of the oviducts and the uterus, and the presence or absence of a urogenital sinus in these two mammals.

**Figure 19.8** **Human female reproductive system.**
Especially compare the anatomy of the oviducts in humans to that of the uterine horns in a pig. In a pig, the fetuses develop in the uterine horns; in a human female, the fetus develops in the body of the uterus.

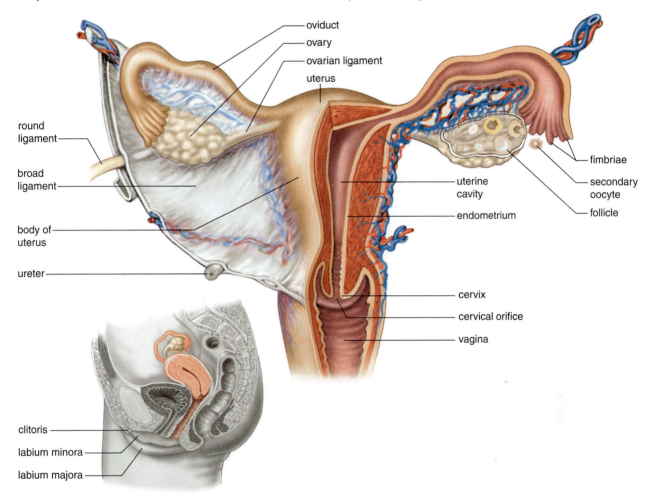

| Table 19.4 | Comparison of Female Fetal Pig with Human Female | |
|---|---|---|
| | **Fetal Pig** | **Human** |
| Oviducts | | |
| Uterus | | |
| Urogenital sinus | | |

# 19.4 Anatomy of Testis and Ovary

Recall that the testes produce sperm (the male gametes) and that the ovaries produce eggs (the female gametes). A testis contains **seminiferous tubules,** where sperm formation takes place, and **interstitial cells** scattered in the spaces between seminiferous tubules. Interstitial cells produce the male sex hormone testosterone. An ovary contains **follicles** in various stages of maturation. Ovarian follicles produce the female sex hormones estrogen and progesterone. One or more follicles complete maturation during each cycle and produce an oocyte.

## Observation: Testis and Ovary

### Testis

1.  Examine a prepared slide of the testis. Note under low power the many circular structures—the seminiferous tubules.
2.  Switch to high power, and observe one tubule in particular. With the help of Figure 19.9, find mature sperm, which look like thin, fine, dark lines, in the middle of the tubule. Interstitial cells are between the tubules.

### Ovary

1.  Examine a prepared slide of an ovary, and refer to Figure 19.10 for help in identifying the structures. The female gonad contains an inner core of loose fibrous tissue. The outer part contains the follicles that produce eggs.
2.  Locate a **primary follicle,** which appears as a circle of cells surrounding a somewhat larger cell.

**Figure 19.9** **Photomicrograph of a seminiferous tubule.**

A testis contains seminiferous tubules separated by interstitial cells. The tubules produce sperm, and the interstitial cells produce male sex hormones.

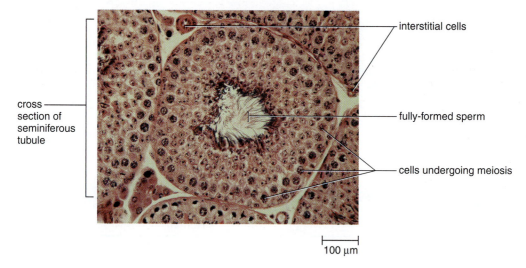

interstitial cells

cross section of seminiferous tubule

fully-formed sperm

cells undergoing meiosis

100 µm

## Figure 19.10 Photomicrograph of ovarian tissue.

An ovary contains follicles in different stages of maturity. A secondary follicle contains a secondary oocyte, which will burst from the ovary during ovulation. A follicle also produces the female sex hormones.

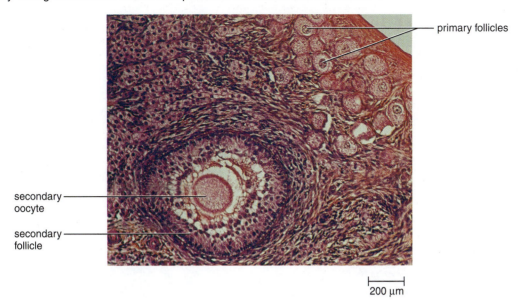

primary follicles

secondary oocyte

secondary follicle

200 µm

3. Find a **secondary follicle,** and switch to high power. Note the **secondary oocyte** (egg), surrounded by numerous cells, to one side of the liquid-filled follicle.
4. Also look for a large, fluid-filled, vesicular (Graafian) follicle, which contains a mature secondary oocyte to one side. This follicle will be next to the outer surface of the ovary because it is the type of follicle that releases the egg during ovulation.
5. Also look for the remains of the corpus luteum, which will look like scar tissue. The corpus luteum develops after the vesicular follicle has released its egg, and then later it deteriorates. Not all slides will contain a vesicular follicle and corpus luteum because they may not have been present when the slide was made.

## Comparison of Reproductive Systems

Complete Table 19.5 to describe the differences between the male and female mammalian reproductive systems.

| Table 19.5 | Comparison of Human Male and Female Reproductive Systems | |
|---|---|---|
| | Male | Female |
| Gonad | | |
| Duct from gonad | | |
| Structure connected to gonad by duct | | |
| Copulatory organ | | |

## Figure 19.11  Internal anatomy of the fetal pig.

Most of the major organs are shown in this photograph. The stomach has been removed. The spleen, gallbladder, and pancreas are not visible.

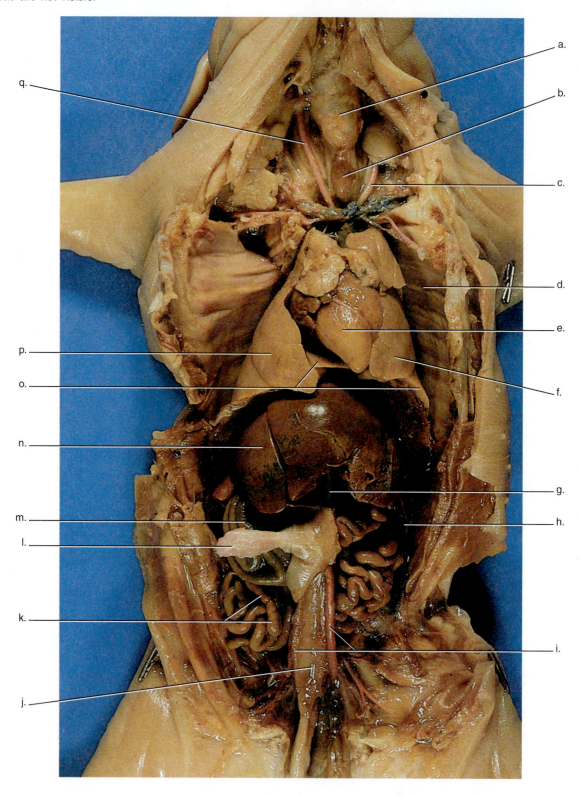

# 19.5 Review of the Respiratory, Digestive, and Cardiovascular Systems

In previous Laboratories, you dissected the respiratory, digestive, and cardiovascular systems of the fetal pig. *Review your knowledge of these systems by reexamining your dissection of the fetal pig and by labeling Figure 19.11.* In this portion of today's lab, you will review each system and examine some organs in more detail. *Do not remove any organs* unless told to do so by your instructor.

## Observation: Respiratory System in Pigs

1. With the help of Figure 19.12, trace the path of air from the nasal passages to the lungs. List the first three organs in the left column and the last three organs in the right column.

   nasal passages

   _____          _____

   _____          _____

   _____          _____

   lungs

2. Make sure you have cut the corners of the mouth as directed in Lab 16, page 211. In the **pharynx,** you should be able to locate the **glottis,** an opening to the _____.
3. If necessary, make a midventral incision in the neck to expose the **larynx.**
4. Clear away the "straplike" muscles covering the **trachea.** Now you should be able to feel the cartilaginous rings that hold the trachea open. Locate the esophagus, which lies below the trachea.
5. If available, observe a slide on display showing a section through the trachea and esophagus. Notice in Figure 19.12 that the air and food pathways cross in the pharynx.

**Figure 19.12  Air and food passages in the fetal pig.**
A probe can pass from the mouth to the larynx to the esophagus.

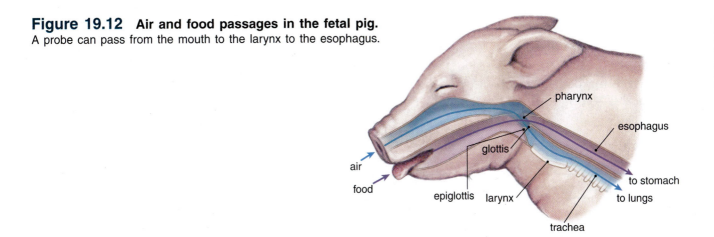

6. Open the pig's mouth, insert a blunt probe into the glottis, and carefully work the probe down through the larynx to the level of the **bronchi.**
7. Observe the **lungs,** and if available, observe a prepared slide of lung tissue.
8. If so directed by your instructor, remove a portion of the trachea, the bronchi, and the lungs, keeping them all in one piece. Place this specimen in a small container of water. Holding the trachea with your forceps, gently but firmly stroke the lung repeatedly with the blunt wooden base of one of your probes. If you work carefully, the alveolar tissue will be fragmented and rubbed away, leaving the branching system of air tubes and blood vessels.

1. Trace the path of food from the mouth to the anus:

   mouth

   _____          _____

   _____          _____

   _____          _____

                          anus

2. Open the **mouth** again, and insert a blunt probe into the esophagus (see Fig. 19.12). Then trace the **esophagus** to the stomach.

3. Open one side of the **stomach,** and examine its interior surface. Does it appear smooth or rough? _____

4. Find the pyloric sphincter, the muscle that surrounds the entrance to the duodenum, the first part of the **small intestine.** Record the length of the small intestine. _____
   If you have not done so before, find the bile duct that empties into the duodenum. The bile duct comes from the _____.

5. Find the **cecum,** a projection where the small intestine enters the large intestine. How does the appearance of the pig's large intestine differ from that of a human?

   _____

6. Carefully cut the mesenteries holding the colon of the **large intestine** in place, and uncoil the large intestine also. Record the length of the large intestine. _____ How does the length of the large intestine compare with that of the small intestine?

   _____

7. Locate again the liver, pancreas, and gallbladder, three accessory organs of digestion.

## Observation: Cardiovascular System in Pigs

### Heart

1. Trace the path of blood through the heart, starting with the vena cava and ending with the aorta. Mention all the chambers of the heart and the valves (see Fig. 17.3).

   To the heart:                    From the lungs:

   vena cava                        _____

   _____                      _____

   _____ valve                _____ valve

   _____                      _____

   _____ valve                _____ valve

   _____                      aorta

   _____

2. Keeping the heart inside the pig, cut the pericardial sac (the tissue that surrounds the heart).
3. Look for and identify the vessels attached to the heart.
4. Section the heart, and look for its four chambers. Remnants of the atrioventricular valves can be seen as thin sheets of whitish tissue attached to fine, white, tendinous strands.
5. With your blunt probe, find the oval opening in the wall between the two atrial chambers. Recall that this is a shunt that allows blood to bypass lung circulation prior to birth.

### Blood Vessels

1. In general, arteries take blood _____ the heart, and veins take blood _____ the heart.
2. Locate the following blood vessels in your pig. State their origin and destination. (For simplicity, use the name of the structure and either the aorta, the anterior vena cava, or the posterior vena cava before or after the name of the structure, as appropriate.)

   a. **coronary artery:** takes blood from the _____ to the _____

   b. **cardiac vein:** takes blood from the _____ to the _____

   c. **carotid artery:** takes blood from the _____ to the _____

   d. **jugular vein:** takes blood from the _____ to the _____

   e. **subclavian artery:** takes blood from the _____ to the _____

   f. **subclavian vein:** takes blood from the _____ to the _____

   g. **renal artery:** takes blood from the _____ to the _____

   h. **renal vein:** takes blood from the _____ to the _____

   i. **iliac artery:** takes blood from the _____ to the _____

   j. **iliac vein:** takes blood from the _____ to the _____

### Hepatic Portal System and Associated Vessels

A **portal system** is a circulatory unit that goes from one capillary bed to another without passing through the heart. For example, in mammals, the mesenteric arteries take blood to the intestines. Thereafter, the hepatic portal vein takes blood from the intestinal capillaries to capillaries in the liver (Fig. 19.13). The significance of this circulatory arrangement is apparent, given the liver's important role in processing and storing materials absorbed from the intestine. The hepatic veins take blood from the liver to the inferior (posterior) vena cava.

1. To find the **hepatic portal vein** in your pig, carefully break the mesenteries in the region of the bile duct. The hepatic portal vein is dorsal to the duct and will not be blue if the latex did not enter it.
2. To see the **hepatic veins,** scrape away the liver substance with the edge of the forceps or the blunt side of the scalpel until all of the soft liver material has been removed and only a mass of cords remains.
3. Identify the **umbilical vein** leading into the liver; the **venous duct,** the main channel through the liver; and the hepatic veins, consisting of three or four vessels from the liver to the posterior vena cava. The great majority of the cords you have exposed consist of branches of veins and bile ducts.

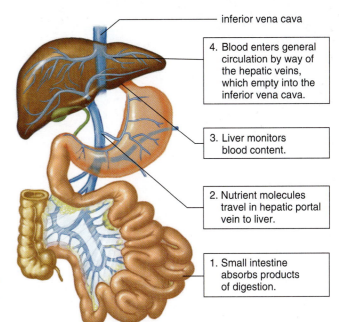

inferior vena cava

4. Blood enters general circulation by way of the hepatic veins, which empty into the inferior vena cava.

3. Liver monitors blood content.

2. Nutrient molecules travel in hepatic portal vein to liver.

1. Small intestine absorbs products of digestion.

**Figure 19.13** **Hepatic portal system.**
The hepatic portal vein takes the products of digestion from the digestive system to the liver, where they are processed. The hepatic veins take blood from the liver to the posterior vena cava.

Basic Mammalian Anatomy II Laboratory 19 **255**

## Storage of Pigs

1. Before leaving the laboratory, place your pig in the plastic bag provided.
2. Expel excess air from the bag, and tie it shut.
3. Write your *name* and *section* on the tag provided, and attach it to the bag. Your instructor will indicate where the bags are to be stored until the next laboratory period.
4. Clean the dissecting tray and tools, and return them to their proper location.
5. Wipe off your goggles.
6. Wash your hands.

## Laboratory Review 19

_____ 1. Which structure in the urinary system carries urine to the bladder?

_____ 2. Which structure in the urinary system receives urine from the bladder?

_____ 3. What is the outermost portion of the kidney?

_____ 4. Where are the testes located in human males?

_____ 5. What is the function of the vas deferens?

_____ 6. What is the function of the prostate gland?

_____ 7. Where are the ovaries located?

_____ 8. What is the function of the uterus?

_____ 9. What is the function of the ovaries?

_____ 10. The vas deferens in males compares with which structure in females?

_____ 11. Which type of mammal, a pig or a human, has uterine horns?

_____ 12. What organ in males is analogous to the vagina?

_____ 13. Where are sperm produced in the testes?

_____ 14. What structure in the ovary contains the developing oocyte?

_____ 15. Name a circulatory pathway that goes from one capillary bed to another without passing through the heart.

### Thought Questions

16. On the basis of anatomy, explain why the urethra is part of both the urinary and reproductive systems in males.

17. Explain the dual function of the vagina.

# 20

# Nervous System and Senses

## Learning Objectives

**20.1 Animal Nervous Systems**
- Identify the three major parts of the sheep brain and state significant functions for each part.
- Compare the brains of various vertebrates in terms of the forebrain, midbrain, and hindbrain.
- Describe the anatomy and physiology of the human spinal cord and spinal nerves.
- Explain the operation of a reflex as an automatic response to a stimulus.

**20.2 Animal Eyes**
- Compare the eyes of various invertebrates, particularly those of flies and squids.
- Identify, locate, and state the functions of the major parts of the human eye.
- Explain how to test accommodation and the blind spot.

**20.3 Animal Ears**
- Compare the tympanum of grasshoppers and the lateral line of fishes to the human ear.
- Identify, locate, and state the functions of the major parts of the human ear.

**20.4 The Senses of Human Skin**
- Describe the anatomy of human skin including its sensory receptors.

**20.5 Animal Chemoreceptors**
- Compare chemoreception in humans to that in flies.

**Figure 20.1  Motor neuron anatomy.**
Neurons are cells specialized to conduct nerve impulses.

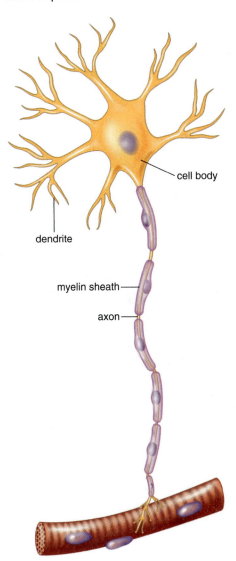

cell body

dendrite

myelin sheath

axon

## Introduction

The vertebrate nervous system consists of the brain, spinal cord, and nerves. Sensory receptors detect changes in environmental stimuli, and nerve impulses move along sensory nerve fibers to the brain and the spinal cord. The brain and spinal cord sum up the data before sending impulses via motor nerve fibers to effectors (muscles and glands) so a response to stimuli is possible. Nervous tissue consists of neurons; whereas the brain and spinal cord contain all parts of neurons, nerves contain only axons (Fig. 20.1).

# 20.1 Animal Nervous Systems

The brain is the enlarged, anterior end of the nerve cord. In vertebrates, the nerve cord is called the spinal cord. The brain contains parts and centers that receive input from, and can command other regions of, the nervous system.

## Sheep Brain

The mammalian brain has many parts, and the sheep brain is often used to study the mammalian brain.

### Observation: Sheep Brain

Examine the exterior and a longitudinal section of a preserved sheep brain or a model of the human brain, and with the help of Figure 20.2, identify the:

1. **Ventricles:** Interconnecting spaces that produce and serve as a reservoir for cerebrospinal fluid, which cushions the brain. Toward the anterior, note the large first lateral ventricle (on one of your transverse sections) and the second lateral ventricle (on your other section). Trace a lateral ventricle to the third and then the fourth ventricles.
2. **Medulla oblongata** (or simply **medulla**): The most posterior portion of the brain stem. It controls internal organs; for example, cardiac and breathing control centers are present in the medulla. Nerve impulses pass from the spinal cord through the medulla to higher brain regions.
3. **Pons:** The ventral, bulblike enlargement on the brain stem. It serves as a passageway for nerve impulses running between the medulla and the higher brain regions.
4. **Midbrain:** Anterior to the pons, the midbrain serves as a relay station for sensory input and motor output. It also contains a reflex center for eye movement.
5. **Diencephalon:** The portion of the brain where the third ventricle is located. The hypothalamus and thalamus also are located here.
6. **Hypothalamus:** Forms the floor of the third ventricle and contains control centers for appetite, body temperature, and water balance. Its primary function is homeostasis. The hypothalamus also has centers for pleasure, reproductive behavior, hostility, and pain.
7. **Thalamus:** Two connected lobes located in the roof of the third ventricle. The thalamus is the highest portion of the brain to receive sensory impulses before the cerebrum. It is believed to control which of these impulses is passed on to the cerebrum. For this reason, the thalamus sometimes is called the "gatekeeper to the cerebrum."
8. **Cerebellum:** Located just posterior to the cerebrum as you observe the brain dorsally, the cerebellum's two lobes make it appear rather like a butterfly. In cross section, the cerebellum has an internal pattern that looks like a tree. The cerebellum coordinates equilibrium and motor activity to produce smooth movements.
9. **Cerebrum:** The most developed area of the brain and responsible for higher mental capabilities. The cerebrum is divided into the right and left **cerebral hemispheres,** joined by the **corpus callosum,** a broad sheet of white matter. The outer portion of the cerebrum is highly convoluted and divided into the following surface lobes:

    a. **Frontal lobe:** Controls motor functions and permits voluntary muscle control. It also is responsible for abilities to think, problem solve, and speak.
    b. **Parietal lobe:** Receives information from sensory receptors located in the skin. It also helps in the understanding of speech. A groove called the **central sulcus** separates the frontal lobe from the parietal lobe.
    c. **Occipital lobe:** Interprets visual input and also combines visual images with other sensory experiences. The optic nerves split and enter opposite sides of the brain at the optic chiasma.
    d. **Temporal lobe:** Has sensory areas for hearing and smelling. The olfactory bulb contains nerve fibers that communicate with the olfactory cells in the nasal passages and take nerve impulses to the temporal lobe.

# Figure 20.2 The sheep brain.

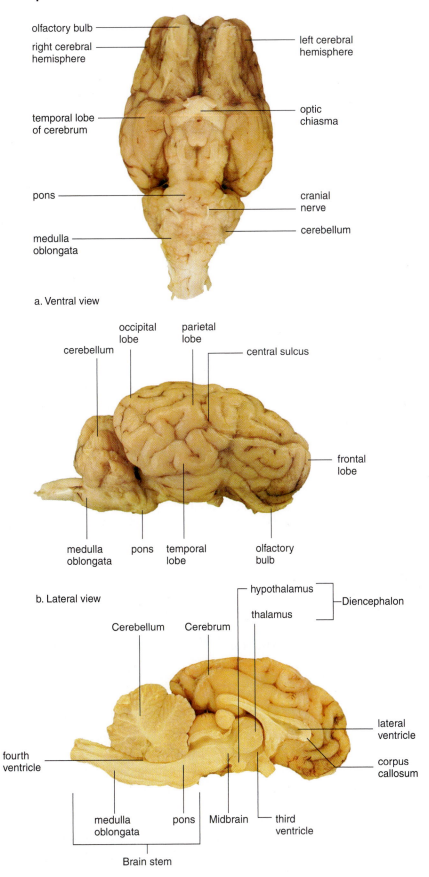

olfactory bulb

right cerebral hemisphere

left cerebral hemisphere

temporal lobe of cerebrum

optic chiasma

pons

cranial nerve

cerebellum

medulla oblongata

a. Ventral view

occipital lobe

parietal lobe

cerebellum

central sulcus

frontal lobe

medulla oblongata

pons

temporal lobe

olfactory bulb

b. Lateral view

hypothalamus

Diencephalon

thalamus

Cerebellum

Cerebrum

lateral ventricle

fourth ventricle

corpus callosum

medulla oblongata

pons

Midbrain

third ventricle

Brain stem

c. Longitudinal cut

## Comparison of Vertebrate Brains

The vertebrate brain has a forebrain, midbrain, and hindbrain. In the earliest vertebrates, the forebrain was largely a center for sense of smell, the midbrain was a center for the sense of vision, and the hindbrain was a center for the sense of hearing and balance. How the functions of these parts changed to accommodate the lifestyles of different vertebrates can be traced.

### Observation: Comparison of Vertebrate Brains

1. Examine the brain of a reptile, bird, and mammal (Fig. 20.3).
2. Identify the following three areas, and for each animal featured.
   a. **Forebrain:** Contains olfactory bulb and cerebrum.
   b. **Midbrain:** Contains optic lobe (and other structures).
   c. **Hindbrain:** Contains cerebellum (and other structures).

In fishes and amphibians, the midbrain is often the most prominent region of the brain because it contains higher control centers. In these animals, the cerebrum largely has an olfactory function. Fishes, birds, and mammals have a well-developed cerebellum, which may be associated with these animals' agility. In reptiles, birds, and mammals, the cerebrum becomes increasingly complex. This may be associated with the cerebrum's increasing control over the rest of the brain and the evolution of areas responsible for thought and reasoning.

**Figure 20.3** **Vertebrate brains.**
*a.* A comparison of reptile, bird, and mammalian brains shows that the forebrain increased in size and complexity among these animals. *b.* The human brain.

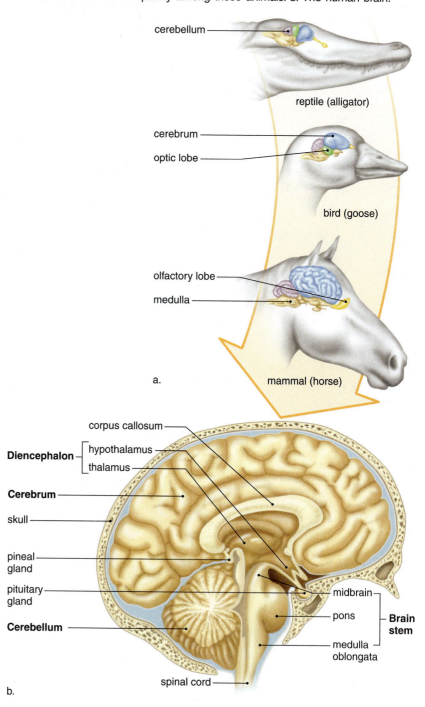

cerebellum

reptile (alligator)

cerebrum
optic lobe

bird (goose)

olfactory lobe
medulla

a.

mammal (horse)

corpus callosum
hypothalamus
Diencephalon
thalamus
Cerebrum
skull
pineal gland
pituitary gland
Cerebellum
midbrain
pons
Brain stem
medulla oblongata
spinal cord
b.

## Anatomy of Spinal Nerves and Spinal Cord

In humans, pairs of spinal nerves are connected to the spinal cord, which lies in the middorsal region of the body protected by the vertebral column. Each spinal nerve contains long fibers of sensory neurons and long fibers of motor neurons. In Figure 20.4, identify the following:

1. **Sensory axon** takes nerve impulses from a sensory receptor to the spinal cord. The cell body of a sensory neuron is in the dorsal-root ganglion. Why is this neuron called a sensory neuron?

   _____

   _____

2. **Interneuron,** which lies completely within the spinal cord. Some interneurons have long fibers and take nerve impulses to and from the brain. The neuron in Figure 20.4 transmits nerve impulses from the sensory neuron to the motor neuron. Why is this neuron called an interneuron?

   _____

   _____

3. **Motor axon** takes nerve impulses from the spinal cord to an effector—in this case, a muscle. Muscle contraction often allows a response to a stimulus. Why is this neuron called a motor

   neuron? _____


**Figure 20.4** **A reflex arc showing the path of a spinal reflex.**
A stimulus (e.g., a pinprick) causes sensory receptors in the skin to generate nerve impulses that travel in sensory axons to the spinal cord. Interneurons integrate data from sensory neurons and then relay signals to motor neurons. Motor axons convey nerve impulses from the spinal cord to a skeletal muscle, which contracts. Movement of the hand away from the pin is the response to the stimulus.

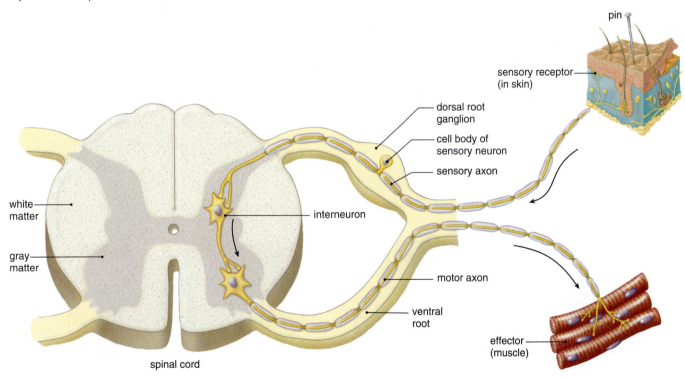

## The Spinal Cord

The spinal cord is a part of the central nervous system. It lies in the middorsal region of the body and is protected by the vertebral column.

### Observation: Spinal Cord

1. Examine a prepared slide of a cross section of the spinal cord under the lowest magnification possible. For example, some microscopes are equipped with a very short scanning objective that enlarges about $3.5\times$, with a total magnification of $35\times$. If neither of these alternatives is available, then observe the slide against a white background with the naked eye.
2. Identify the following with the help of Figure 20.5:
   a. **Gray matter:** A central, butterfly-shaped area composed of masses of short nerve fibers, interneurons, and motor neuron cell bodies.
   b. **White matter:** Masses of long fibers that lie outside the gray matter and carry impulses up and down the spinal cord. In living animals, white matter appears white because an insulating myelin sheath surrounds long fibers.

**Figure 20.5** The spinal cord.
Photomicrograph of spinal cord cross section.

central canal

gray matter

white matter

## Spinal Reflexes

A **reflex** is an involuntary and predictable response to a given stimulus. When you touch a sharp tack, you immediately withdraw your hand (see Fig. 20.4). When a spinal reflex occurs, a sensory receptor is stimulated and generates nerve impulses that pass along the three neurons mentioned earlier—the sensory neurons, interneurons, and motor neurons—until the effector responds.

### Experimental Procedure: Spinal Reflexes

Although many reflexes occur in the body, only tendon reflexes are investigated in this Experimental Procedure. Two easily tested tendon reflexes involve the **Achilles** and **patellar tendons.** When these tendons are mechanically stimulated by tapping with a reflex hammer (Fig. 20.6). Sensory neurons

**Figure 20.6** Two human reflexes.
The quick response when either the a. Achilles tendon or the b. patellar tendon is stimulated by tapping with a rubber hammer indicates that a reflex has occurred.

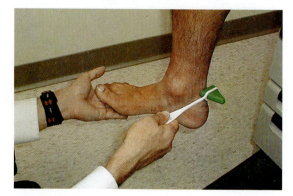

a. Ankle (Achilles) reflex

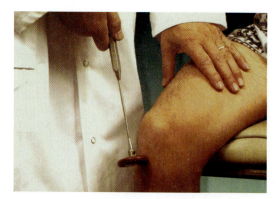

b. Knee-jerk (patellar) reflex

in the tendon transmit impulses through the reflex arc to the spinal cord and then back through motor neurons to the muscle attached to the stimulated tendon. The muscle contracts and tugs on the tendon, causing movement of a bone opposite the joint.

### Ankle (Achilles) Reflex

1. Have the subject sit on the table so that the leg hangs free.
2. Tap the subject's Achilles tendon at the ankle with a reflex hammer.
3. Which way does the foot move? Does it extend (move away from the knee) or flex (move toward the knee)? _____

### Knee-Jerk (Patellar) Reflex

1. Have the subject sit on the table so that the leg hangs free.
2. Sharply tap the patellar tendon just below the patella (kneecap) with a reflex hammer.
3. In this relaxed state, does the leg flex (move toward the buttocks) or extend (move away from the buttocks)? _____

## 20.2 Animal Eyes

The eye is a special sense organ for detecting light rays in the environment.

### Anatomy of Invertebrate Eyes

Arthropods have **compound eyes** composed of many independent visual units, each of which has its own lens (Fig. 20.7a). Each unit "sees" a separate portion of the object. How well the brain combines this information is not known.

A chambered nautilus has a camera type of eye. A single lens focuses an image of the visual field on the photoreceptors, packed closely together (Fig. 20.7b).

### Observation: Invertebrate Eyes

1. Examine the demonstration slide of a compound eye set up under a binocular dissecting microscope or examine a model of a compound eye.
2. Examine the eyes of any other invertebrates on display.

### Figure 20.7 The compound eye compared to the camera-type eye.

*a.* A fly has a compound eye. Each visual unit of a compound eye has a cornea and lens that focuses light onto photoreceptor cells. These cells generate nerve impulses transmitted to the brain, where interpretation produces a mosaic image. *b.* A cephalopod, such as a chambered nautilus and a squid, have a camera-type eye similar to those of vertebrates.

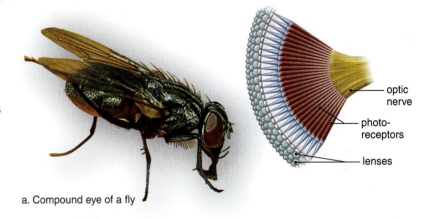

a. Compound eye of a fly

optic nerve
photo-receptors
lenses

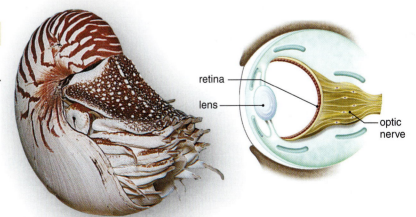

b. Camera-type eye of a chambered nautilus

retina
lens
optic nerve

## Anatomy of the Human Eye

The anatomy of the human eye allows light rays to enter the eye and strike the **rod cells** and **cone cells,** the photoreceptors for sight. The rods and cones generate nerve impulses that go to the brain via the optic nerve.

### Observation: Human Eye

1. Examine a human eye model, and identify the structures listed in Table 20.1 and depicted in Figure 20.8.
2. Trace the path of light from outside the eye to the retina.

_____

_____

3. Specifically, what are the receptors for sight, and where are they located in the eye?

_____

4. What structure takes nerve impulses to the brain from the rod cells and cone cells?

_____

**Figure 20.8** **Anatomy of the human eye.**
The sensory receptors for vision are the rod cells and cone cells present in the retina of the eye.

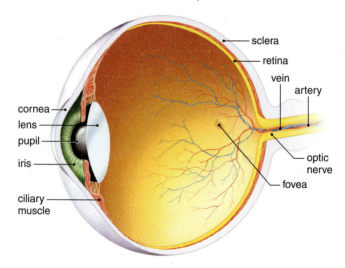

## Physiology of the Human Eye

When we look at an object, light rays pass through the cornea, and as they do, they bend and converge. Further bending occurs as the rays pass through the lens. The shape of the lens changes according to whether we are looking at a far or near object. When looking at a near object, the lens rounds so that the light rays are bent a great deal to bring all rays to a focus on the retina. You will be measuring the ability of your lens to round up (accommodate) for viewing a near object.

Where the optic nerve pierces the wall of the eyeball and, therefore, the retina, there are no rod cells and cone cells, and therefore vision is impossible. You will be finding the blind spot in an Experimental Procedure.

## Table 20.1 — Parts of the Human Eye

| Part | Location | Function |
|------|----------|----------|
| Sclera | Outer layer of the eye | Protects and supports eyeball |
| Cornea | Transparent portion of sclera | Refracts light rays |
| Choroid | Middle layer of the eye | Absorbs stray light rays |
| Retina | Inner layer of the eye | Contains receptors for sight |
| Rod cells | In retina | Make black-and-white vision possible |
| Cone cells | Concentrated in fovea centralis | Make color vision possible |
| Fovea centralis | Special region of retina | Makes acute vision possible |
| Lens | Interior of eye between cavities | Refracts and focuses light rays |
| Ciliary body | Extension from choroid | Holds lens in place; functions in accommodation |
| Iris | More anterior extension of choroid | Regulates light entrance |
| Pupil | Opening in middle of iris | Admits light |
| Humors (aqueous and vitreous) | Fluid media in anterior and posterior compartments, respectively, of eye | Transmit and refract light rays; support the eyeball |
| Optic nerve | Extension from posterior of eye | Transmits impulses to brain |

### Experimental Procedure: Accommodation of the Eye

When the eye accommodates to see objects at different distances, the shape of the lens changes. The lens shape is controlled by the ciliary muscles attached to it. When you are looking at a distant object, the lens is in a flattened state. When you are looking at a closer object, the lens becomes more rounded. The elasticity of the lens determines how well the eye can accommodate. Lens elasticity decreases with increasing age, a condition called **presbyopia.** Because of presbyopia, many older people need bifocals to see near objects.

This Experimental Procedure requires a laboratory partner. It tests accommodation of either your left or right eye.

1. Hold a pencil upright by the eraser and at arm's length in front of whichever of your eyes you are testing (Fig. 20.9).
2. Close the opposite eye.
3. Move the pencil from arm's length toward your eye.
4. Focus on the end of the pencil.
5. Move the pencil toward you until the end is out of focus. Measure the distance (in centimeters) between the pencil and your eye. _____ cm
6. At what distance can your eye no longer accommodate for distance? _____ cm

**Figure 20.9** **Accommodation.**
The lens will round up to see the end of the pencil.

7. If you wear glasses, repeat this experiment without your glasses, and note the accommodation distance of your eye without glasses. _____ cm (Contact lens wearers need not make these determinations, and they should write the words *contact lens* in this blank.)

8. The "younger" lens can easily accommodate for closer distances. The nearest point at which the end of the pencil can be clearly seen is called the **near point.** The more elastic the lens, the "younger" the eye (Table 20.2). How "old" is the eye you tested? _____

| Table 20.2 | Near Point and Age Correlation | | | | | |
|---|---|---|---|---|---|---|
| Age (Years) | 10 | 20 | 30 | 40 | 50 | 60 |
| Near Point (cm) | 9 | 10 | 13 | 18 | 50 | 83 |

## Figure 20.10 Blind spot.

This dark circle (or cross) will disappear at one location because there are no rod cells or cone cells at each eye's blind spot, where vision does not occur.

### Experimental Procedure: Blind Spot of the Eye

The **blind spot** occurs where the optic nerves exit the retina (see Fig. 20.8). No vision is possible at this location because of the absence of rods and cones.

This Experimental Procedure requires a laboratory partner. Figure 20.10 shows a small circle and a cross several centimeters apart.

#### Left Eye

1. Hold Figure 20.10 approximately 30 cm from your eyes. If you wear glasses, keep them on.
2. Close your right eye.
3. Stare only at the cross with your left eye. You should also be able to see the circle in the same field of vision. Slowly move the paper toward you until the circle disappears.
4. Repeat the procedure as many times as needed to find the blind spot.
5. Then slowly move the paper closer to your eyes until the circle reappears. Because only your left eye is open, you have found the blind spot of your left eye.
6. Measure the distance (with your partner's help) from your eye to the paper when the circle first disappeared. _____ cm, left eye

#### Right Eye

1. Hold Figure 20.10 approximately 30 cm from your eyes. If you wear glasses, keep them on.
2. Close your left eye.
3. Stare only at the circle with your right eye. You should also be able to see the cross in the same field of vision. Slowly move the paper toward you until the cross disappears.
4. Repeat the procedure as many times as needed to find the blind spot.
5. Then slowly move the paper closer to your eyes until the cross reappears. Because only your right eye is open, you have found the blind spot of your right eye.
6. Measure the distance (with your partner's help) from your eye to the paper when the cross first disappeared. _____ cm, right eye

# 20.3 Animal Ears

Ears contain specialized receptors for detecting sound waves in the environment. They also often function as organs of balance.

## Anatomy of Invertebrate Ears

Among invertebrates, only certain arthropod groups—crustaceans, spiders, and insects—have receptors for detecting sound waves. The invertebrate ear usually has a simple design: a pair of air pockets enclosed by a tympanum that passes sound waves to sensory neurons.

### Observation: An Invertebrate Ear

Examine the preserved grasshopper on display, and with the help of Figure 20.11a, locate the tympanum. The tympanum covers an internal air sac that allows the tympanum to vibrate when struck by sound waves. Sensory neurons attached to the tympanum are stimulated directly by the vibration.

## Anatomy of the Human Ear

The tetrapod vertebrate ear may have evolved from the fishes. Fishes have a lateral line, a series of hair cells with cilia embedded in a mass of gelatinous material that detects water currents and pressure waves from nearby objects (Fig. 20.11b). Fishes also have **otic vesicles** that mainly function as organs of balance. In contrast to the "ear" of fishes, the anatomy of the human ear allows sound waves to be received, amplified, and detected by receptors. The human ear also contains receptors for balance.

**Figure 20.11** **Evolution of the human ear.**
a. A few invertebrates have ears such as that of the grasshopper. b. The lateral line of fishes is not for hearing or balance. It is for knowing the location of other fishes.

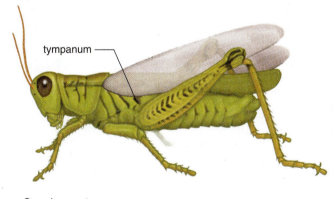

a. Grasshopper tympanum

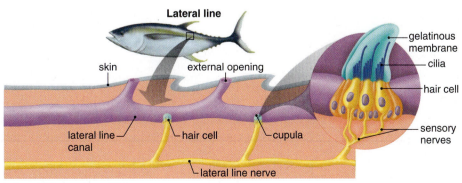

b. Lateral line system of fishes

Nervous System and Senses  Laboratory 20  **267**

Examine a human ear model, and identify the structures depicted in Figure 20.12 and listed in Table 20.3.

## Figure 20.12   Human ear.

*a.* The outer ear collects sound waves, the middle ear amplifies sound waves, and the inner ear (cochlea) contains the sensory receptors for hearing. *b.* The inner ear contains the cochlea, and also the semicircular canals and the vestibule.

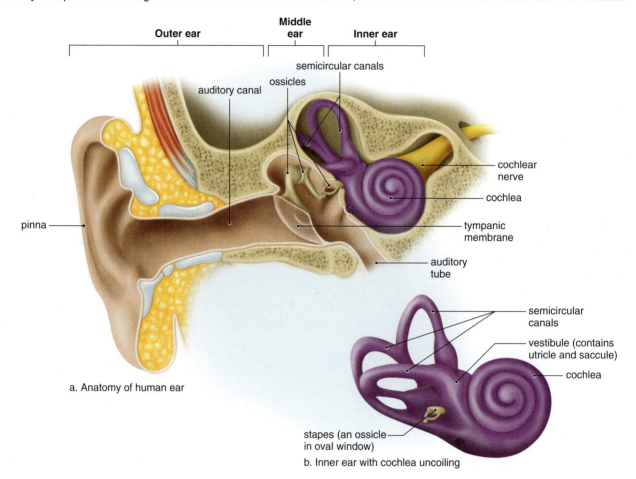

a. Anatomy of human ear

b. Inner ear with cochlea uncoiling

| Table 20.3 | Parts of the Human Ear | | |
|---|---|---|---|
| **Part** | **Medium** | **Function** | **Mechanoreceptor** |
| *Outer ear* | Air | | |
| Pinna | | Collects sound waves | — |
| Auditory canal | | Filters air | — |
| *Middle ear* | Air | | |
| Tympanic membrane and ossicles | | Amplify sound waves | — |
| Auditory tube | | Equalizes air pressure | — |
| *Inner ear* | Fluid | | |
| Semicircular canals | | Rotational equilibrium | Stereocilia embedded in cupula |
| Vestibule (contains utricle and saccule) | | Gravitational equilibrium | Stereocilia embedded in otolithic membrane |
| Cochlea (spiral organ) | | Hearing | Stereocilia embedded in tectorial membrane |

## Physiology of the Human Ear

The process of hearing begins when sound waves enter the auditory canal. The sound waves are detected by the tympanic membrane and amplified by the ossicles (bones). The last ossicle, the stapes, strikes the oval window of the cochlea. As a result of the movement of the fluid within the cochlea, the hair cells in the spiral organ (organ of Corti) are stimulated. The hair cells are sensory receptors that cause nerve impulses to be conducted in the cochlear (auditory) nerve to the brain.

### Experimental Procedure: Locating Sound

Humans locate the direction of sound according to how fast it is detected by either or both ears. A difference in the hearing ability of the two ears can lead to a mistaken judgment about the direction of sound. Both you and a laboratory partner should perform this Experimental Procedure on each other. Enter the data for *your* ears, not your partner's ears, in your laboratory manual.

1. Ask *the subject* to be seated, with eyes closed.
2. Then strike a tuning fork or rap two spoons together at the five locations listed in number 4. Use a random order.
3. Ask the subject to give the exact location of the sound in relation to his or her head.
4. Record the subject's perceptions when the sound is

    a. directly below and behind the head _____

    _____

    b. directly behind the head _____

    c. directly above the head _____

    _____

    d. directly in front of the face _____

    _____

    e. to the side of the head _____

    _____

5. Is there an apparent difference in hearing between your two ears? _____

    _____

## 20.4 The Senses of Human Skin

The receptors in human skin are sensitive to touch, pain, temperature, and pressure. There are individual receptors for each of these various stimuli.

With the help of a model of human skin (Fig. 20.13), locate the following areas or structures, and describe the location of each:

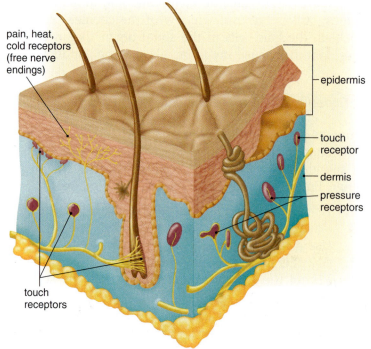

1.  Subcutaneous layer _____
2.  Adipose tissue _____
3.  Dermis _____
4.  Epidermis _____
5.  Hair follicle and hair _____
6.  Oil gland _____
7.  Sweat gland _____
8.  Sensory receptors _____

**Figure 20.13** **Cutaneous receptors.**
Numerous receptors are in the skin. Free nerve endings in the epidermis detect pain, heat, and cold. Various touch and pressure receptors are in the dermis. Sensory receptors have been colored red for easy viewing.

## Sense of Touch

The dermis of the skin contains touch receptors whose concentration differs in various parts of the body.

### Experimental Procedure: Sense of Touch

You will need a laboratory partner to perform this Experimental Procedure. Enter *your* data, not the data of your partner, in your laboratory manual.

1.  Ask *the subject* to be seated, with eyes closed.
2.  Then test the subject's ability to discriminate between the two points of a hairpin or a pair of scissors at the four locations listed in number 5.
3.  Hold the points of the hairpin or scissors on the given skin area, with both of the points simultaneously and gently touching the subject.
4.  Ask the subject whether the experience involves one or two touch sensations.
5.  Record the shortest distance between the hairpin or scissor points for a two-point discrimination in the following areas:

    a.  Forearm _____ mm

    b.  Back of the neck _____ mm

    c.  Index finger _____ mm

    d.  Back of the hand _____ mm

6.  Which of these areas apparently contains the greatest density of touch receptors? _____
    Why is this useful? _____

## Sense of Heat and Cold

Temperature receptors respond to a change in temperature.

### Experimental Procedure: Sense of Heat and Cold

1. Obtain three 1,000 ml beakers, and fill one with *ice water,* one with *tap water* at room temperature, and one with *warm water* (45–50°C).
2. Immerse your left hand in the ice water beaker and your right hand in the warm water beaker for 20 seconds.
3. Then place both hands in the beaker with room-temperature tap water.
4. Record the sensation in the right and left hands.
   a. Right hand _____
   b. Left hand _____

## 20.5 Animal Chemoreceptors

Taste buds located on the tongue and olfactory receptors located in the nose are **chemoreceptors** in humans. Most other animals also have receptors that respond to chemicals in the same manner. For example, flies have chemoreceptors on their legs. When flies are ready to feed, chemoreceptors signal the lowering of the **proboscis,** a tubelike feeding structure (Fig. 20.14).

### Experimental Procedure: Chemoreceptors

1. Obtain a styrofoam block that contains rods with flies attached.
2. Obtain a chemplate, and fill a series of wells with *sugar solutions* of increasing concentrations.
3. Fill one well with water.
4. Give the fly a drink of water before you begin, because a thirsty fly will react the same to water as a hungry fly will to food (sugar).
5. Now hold the fly so that its forelegs come into contact with the sugar solutions, as seen in Figure 20.14b, starting with the least concentrated. *Do not allow the fly to drink.*
6. Rinse the fly's legs between tests, using a wash bottle. (Collect the water in a beaker.) Place an X on the appropriate line whenever your fly lowers its proboscis to feed.

| Sugar | 0.05M | 0.1M | 0.2M | 0.4M | 0.8M |
|---|---|---|---|---|---|
| Sucrose | _____ | _____ | _____ | _____ | _____ |

7. At what sugar concentration does the fly attempt to feed by lowering its proboscis? _____

**Figure 20.14** **Chemoreception in flies.**
Flies lower the proboscis to feed only after the chemoreceptors on their feet respond to an edible substance.

a. Proboscis retracted

b. Fly attached to rod lowered toward food until feet touch the surface

c. Proboscis fully extended

_____  1. What portion of the brain is largest in humans?

_____  2. What portion of the brain controls muscular coordination?

_____  3. What is the most inferior portion of the brain stem?

_____  4. What structures protect the spinal cord?

_____  5. Are motor neuron cell bodies located in the gray or white matter of the spinal cord?

_____  6. What type of neuron is found completely within the CNS?

_____  7. What part of the eye contains the sensory receptors for sight?

_____  8. Where on the retina is the blind spot located?

_____  9. What kind of an eye do arthropods have?

_____  10. What part of the ear contains the sensory receptors for hearing?

_____  11. Where in relation to the head is it most difficult to detect the location of a sound?

_____  12. In which portion of the ear are the malleus, incus, and stapes located?

_____  13. What layer of the skin contains sensory receptors for touch?

_____  14. Are touch receptors distributed evenly or unevenly in the skin?

_____  15. What senses are dependent on chemoreceptors?

_____  16. The human ear may have evolved from the _____ of fishes.

### Thought Questions

17. Trace the path of light in the human eye through each structure or compartment—from the exterior to the retina. How do nerve impulses from the retina reach the brain?

18. Trace the path of sound waves in the human ear—from the tympanic membrane to the sensory receptors for hearing.

# 21

# Effects of Pollution on Ecosystems

## Introduction

**Pollutants** are substances added to the environment, particularly by human activities, that lead to undesirable effects for all living things. In this laboratory, we will examine the effect of two types of pollutants—acid deposition and thermal pollution—on the well being of organisms and hypothesize how these pollutants might affect an ecological pyramid, such as the one shown in Figure 21.1.

We will also explore how ecosystem interactions can change when excess nutrients are added to a system. Over time, bodies of water undergo a natural enrichment process called **eutrophication.** Eutrophication leads to overgrowth that can eventually cause a pond or lake to fill in and disappear. Sometimes, human activities add excess nutrients to bodies of water, and this is called *cultural eutrophication.* Cultural eutrophication leads to lack of oxygen and thus, death of organisms.

**Figure 21.1  Ecological pyramid.**
The biomass at each succeeding trophic (feeding) level in an ecosystem decreases. At the right are examples of organisms at these trophic levels. They form a food chain in which hawks eat other birds, which eat caterpillars, which eat green plants.

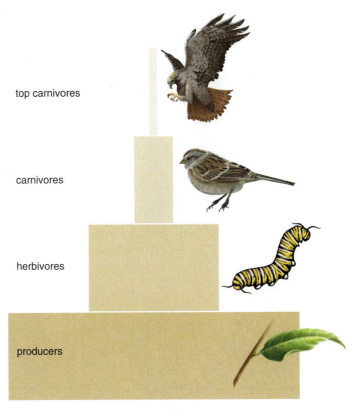

top carnivores

carnivores

herbivores

producers

# 21.1 Studying the Effects of Pollutants

We are going to study the effects of pollution by observing its effects on hay infusion cultures, on seed germination, and on an animal called *Gammarus*.

## Study of Hay Infusion Cultures

A hay infusion culture (hay soaked in water) contains various microscopic organisms, such as those depicted in Figure 21.2. What do you predict will happen to organisms in a hay infusion culture when conditions are acidic or when overenrichment occurs? _____

_____

_____

   When nutrients produced by human activities enter a body of water, the algae overpopulate. Zooplankton are unable to significantly reduce the algal population, and bacteria use all the available oxygen when decomposing them. What do you predict will happen next? _____

_____

### Experimental Procedure: Study of Hay Infusion Cultures

The following four hay infusion cultures have been provided:

1. **Control culture:** This hay infusion culture simulates the optimum conditions for normal growth.
2. **Enriched culture:** Same as control culture, but with more inorganic nutrients for the growth of algae.
3. **Oxygen-deprived culture:** Same as control culture, but with minimal oxygen to test the effect of lack of oxygen on organisms.
4. **Acidic culture:** Same as control culture, except that it is adjusted to pH 4 with sulfuric acid ($H_2SO_4$). This simulates the effect of acid deposition (e.g., acid rain) on organisms.

**Figure 21.2** **Microorganisms in hay infusion cultures.**

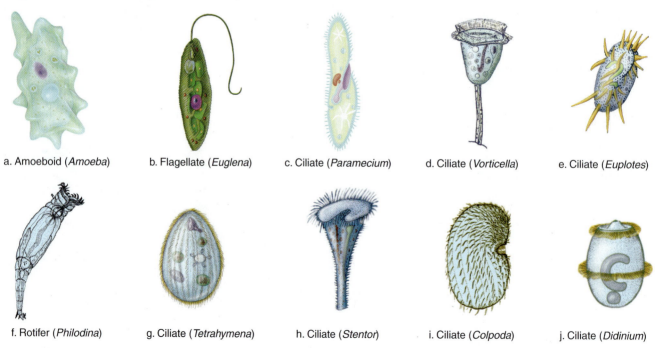

a. Amoeboid (*Amoeba*)   b. Flagellate (*Euglena*)   c. Ciliate (*Paramecium*)   d. Ciliate (*Vorticella*)   e. Ciliate (*Euplotes*)

f. Rotifer (*Philodina*)   g. Ciliate (*Tetrahymena*)   h. Ciliate (*Stentor*)   i. Ciliate (*Colpoda*)   j. Ciliate (*Didinium*)

Examine each of these cultures by preparing a wet mount. Record in Table 21.1 the diversity of life that you observe and the relative quantity of organisms.

| Table 21.1 | Hay Infusion Cultures | | |
| --- | --- | --- | --- |
| Wet Mount | Type of Culture | Diversity of Life (List Organisms) | Relative Quantity of Organisms (High, Medium, or Low) |
| 1 | Control | | |
| 2 | Enriched | | |
| 3 | Oxygen-deprived | | |
| 4 | Acidic | | |

## Conclusions

- How might these observations of the hay infusion experiment relate to real ecosystems? _____

- What are the potential consequences of acid deposition on plant populations that reproduce by seeds? _____

## Study of Seed Germination

Just like animals, plant seeds depend on proper conditions of temperature, light, and moisture to germinate, grow, and reproduce. Many pollutants affect seeds' ability to germinate.

### Experimental Procedure: Study of Seed Germination

Observe these petri dishes, and record your observations in Table 21.2.

1. **Control petri dish:** The seeds in this petri dish have been watered with a control solution having a neutral pH.
2. **Acidic petri dish:** The seeds in this petri dish have been watered with an acidic solution having a pH of 4 to simulate acid rain.

| Table 21.2 | Seed Germination | | |
| --- | --- | --- | --- |
| Petri Dish | Type of Solution | pH | Observations |
| 1 | Control | | |
| 2 | Acidic | | |

## Study of *Gammarus*

We will study the effect of thermal pollution and acid rain on a small crustacean called *Gammarus*, found in ponds and streams (Fig. 21.3).

## Control Culture

1. Add 25 ml of spring water to a container.

2. Measure the pH of the spring water with a pH meter or litmus paper, and record it here: _____.
   Add four *Gammarus* to the container.

3. Observe the behavior of *Gammarus* for 10 to 15 minutes, and then answer the following
   questions. Retain these *Gammarus* for subsequent exercises.

   • Where do the *Gammarus* spend their time in the container? _____
     _____

   • How do they spend their time? _____

   • What percentage of their time is spent moving? _____

   • Do they use all their legs in swimming? _____

   • Which legs are used in jumping and climbing? _____

   • Do *Gammarus* avoid each other? _____

   • What do *Gammarus* do when they "bump" into each other? _____

## Thermal Pollution

1. To simulate thermal pollution, boil 100 ml of
   *water,* and then allow it to cool to 31°C.

2. Put 25 ml of this 31°C water in a beaker, and
   add two *Gammarus.*

3. Observe the behavior of *Gammarus,* and answer
   the following question:

   • What is the effect of thermal pollution on
     the behavior of *Gammarus?*

     _____

     _____

## Acid Pollution

1. Put 25 ml of *acidic spring water* (adjusted to
   pH 4) in a beaker, and add two *Gammarus.*

2. Observe the animals' behavior, and answer
   the following questions:

   • What is the difference between the pH of
     the control culture and the pH of the acidic

     culture? _____

   • Compare the behavior of *Gammarus* in the
     control culture with its behavior in the acidic
     culture. _____

     _____

**Figure 21.3** *Gammarus.*

*Gammarus* is a type of crustacean, a subphylum that
also includes shrimp.

## Conclusion

   • Pollution causes changes in the behavior of *Gammarus.* What consequences might these
     behavioral changes have in a natural food chain? _____
     _____

# 21.2 Studying the Effects of Cultural Eutrophication

*Chlorella,* the alga used in this study, is considered to be representative of **phytoplankton** (free-floating algae) in bodies of fresh water. The crustacean *Daphnia* is a zooplankton that feeds on *Chlorella.* First, you will observe how *Daphnia* feeds, and then you will determine the extent to which *Daphnia* could keep cultural eutrophication from occurring in a hypothetical example. Keep in mind that this case study is an oversimplification of a generally complex problem.

## Observation: Daphnia *Feeding*

1. Place a small pool of petroleum jelly in the center of a small petri dish.
2. Use a dropper to take a *Daphnia* from the stock culture, place it on its back (covered by water) in the petroleum jelly, and observe it under the binocular dissecting microscope.
3. Note the clamlike carapace and the legs waving rapidly as the *Daphnia* filters the water.
4. Add a drop of *carmine solution,* and observe how the *Daphnia* filters the "food" from the water and passes it through the gut. The gut is more visible if you push the animal onto its side. In this position, you may also observe the heart beating in the region above the gut and just behind the head.
5. Allow the *Daphnia* to filter-feed for up to 30 minutes, and observe the progress of the carmine particles through the gut. Does the carmine travel completely through the gut in 30 minutes?

_____

## Experimental Procedure: Daphnia *Feeding on* Chlorella

This exercise requires the use of a spectrophotometer. Absorbance will be a measure of the algal population level; the greater the number of algal cells, the greater the absorbance. The higher the absorbance, the greater the amount of light absorbed and *not* passed through the solution.

1. Obtain two spectrophotometer tubes (cuvettes) and a Pasteur pipette.
2. Fill one of the cuvettes with distilled water, and use it to zero the spectrophotometer. Save this tube for number 6.
3. Use the Pasteur pipette to fill the second cuvette with *Chlorella.* Gently aspirate and expel the sample several times (without creating bubbles) to give a uniform dispersion of the algae.
4. Add ten hungry *Daphnia,* and following your instructor's directions, immediately measure the absorbance with the spectrophotometer. If a *Daphnia* swims through the beam of light, a strong deflection should occur; do not use any such higher readings—instead, use the lower reading for the absorbance. Record your reading in the first column of Table 21.3.
5. Remove the cuvette with the *Daphnia* to a safe place in a test-tube rack. Allow the *Daphnia* to feed for 30 minutes.
6. Rezero the spectrophotometer with the distilled water cuvette.
7. Measure the absorbance of the experimental cuvette again. Record your reading in the second column of Table 21.3, and explain your results in the third column.

| Table 21.3 | Spectrophotometer Data of *Daphnia* Feeding on *Chlorella* | |
|---|---|---|
| **Absorbance Before Feeding** | **Absorbance After Feeding** | **Explanation** |
| | | |

## Experimental Procedure: Case Study in Cultural Eutrophication

The following problem will test your understanding of the value of a single species—in this case, *Daphnia*. Please realize that this is an oversimplification of a generally complex problem.

1. Assume that developers want to build condominium units on the shores of Silver Lake. Homeowners in the area have asked the regional council to determine how many units can be built without altering the nature of the lake. As a member of the council, you have been given the following information:

   The present population of *Daphnia*, 10 animals/liter, presently filters 24% of the lake per day, meaning that it removes this percentage of the algal population per day. This is sufficient to keep the lake essentially clear. Predation—the eating of the algae—will allow the *Daphnia* population to increase to no more than 50 animals/liter. Therefore, 50 *Daphnia*/liter will be available for feeding on the increased number of algae that would result from building the condominiums.

   Using this information, complete Table 21.4.

| Table 21.4 | *Daphnia* Filtering |
|---|---|
| **Number of *Daphnia*/Liter** | **Percent of Lake Filtered** |
| 10 | 24% |
| 50 | |

2. The sewage system of the condominiums will add nutrients to the lake. Phosphorus output will be 1 kg per day for every ten condominiums. This will cause a 30% increase in the algal population. Using this information, complete Table 21.5.

| Table 21.5 | Cultural Eutrophication | |
|---|---|---|
| **Number of Condominiums** | **Phosphorus Added** | **Increase in Algal Population** |
| 10 | 1 kg | 30% |
| 20 | | |
| 30 | | |
| 40 | | |
| 50 | | |

### Conclusions

- Assume that phosphorus is the only nutrient that will cause an increase in the algal population and that *Daphnia* is the only type of zooplankton available to feed on the algae. How many condominiums would you allow the developer to build? _____

- What other possible impacts could condominium construction have on the condition of the lake? _____

_____  1. Any effect on seeds would typify an effect on what trophic level in an ecosystem?

_____  2. Any effect on *Gammarus* would typify an effect on what type of population in an ecosystem?

_____  3. Give an example of acid deposition.

_____  4. What type of pollution results when water from rivers and ponds is used to cool industrial processes?

_____  5. In your experiment, did you add acid or base to adjust the hay infusion culture to pH 4?

_____  6. What condition does acid deposition cause that can be harmful to organisms?

_____  7. Cultural eutrophication begins with an excess of what type of substances?

_____  8. Overenrichment causes which types of populations to increase in size beyond the ordinary?

_____  9. In the case study, what factor caused the producer population (*Chlorella*) to increase?

_____  10. What are free-floating algae called? Why?

## Thought Questions

11. Does pollution affect all living things, including humans? How?

12. When pollutants enter the environment, they have far-ranging effects. Give an example from this laboratory.

# A

# Metric System

| Unit and Abbreviation | Metric Equivalent | Approximate English-to-Metric Equivalents | Units of Temperature |
|---|---|---|---|

**Length**

nanometer (nm) = $10^{-9}$ m ($10^{-3}$ μm)
micrometer (μm) = $10^{-6}$ m ($10^{-3}$ mm)
millimeter (mm) = 0.001 ($10^{-3}$) m
centimeter (cm) = 0.01 ($10^{-2}$) m    1 inch = 2.54 cm
   1 foot = 30.5 cm
   1 foot = 0.30 m
meter (m) = 100 ($10^2$) cm    1 yard = 0.91 m
= 1,000 mm
kilometer (km) = 1,000 ($10^3$) m    1 mi = 1.6 km

**Weight (mass)**

nanogram (ng) = $10^{-9}$ g
microgram (μg) = $10^{-6}$ g
milligram (mg) = $10^{-3}$ g
gram (g) = 1,000 mg    1 ounce = 28.3 g
   1 pound = 454 g
   = 0.45 kg
kilogram (kg) = 1,000 ($10^3$) g
metric ton (t) = 1,000 kg    1 ton = 0.91 t

**Volume**

microliter (μl) = $10^{-6}$ l ($10^{-3}$ ml)
milliliter (ml) = $10^{-3}$ liter
   = 1 cm$^3$ (cc)    1 tsp = 5 ml
   = 1,000 mm$^3$    1 fl oz = 30 ml
liter (l) = 1,000 ml    1 pint = 0.47 liter
   1 quart = 0.95 liter
   1 gallon = 3.79 liter
kiloliter (kl) = 1,000 liter

**Temperature scale (°F / °C):** 212°/210, 160°/160, 134°/131°/130, 105.8°/98.6°/100, 56.66°/60, 32°/30; °C side: 100–100°, 70–71°, 57°, 41°/37°, 13.7°, 0–0°

**Common Temperatures**

| °C | °F | |
|---|---|---|
| 100 | 212 | Water boils at standard temperature and pressure. |
| 71 | 160 | Flash pasteurization of milk |
| 57 | 134 | Highest recorded temperature in the United States, Death Valley, July 10, 1913 |
| 41 | 105.8 | Average body temperature of a marathon runner in hot weather |
| 37 | 98.6 | Human body temperature |
| 13.7 | 56.66 | Human survival is still possible at this temperature. |
| 0 | 32.0 | Water freezes at standard temperature and pressure. |

To convert temperature scales:

$$°C = \frac{5(°F-32)}{9}$$

$$°F = \frac{9°C}{5} + 32$$

Name _____ Section _____ Date _____

## Practical Examination Answer Sheet

1. _____     26. _____
2. _____     27. _____
3. _____     28. _____
4. _____     29. _____
5. _____     30. _____
6. _____     31. _____
7. _____     32. _____
8. _____     33. _____
9. _____     34. _____
10. _____     35. _____
11. _____     36. _____
12. _____     37. _____
13. _____     38. _____
14. _____     39. _____
15. _____     40. _____
16. _____     41. _____
17. _____     42. _____
18. _____     43. _____
19. _____     44. _____
20. _____     45. _____
21. _____     46. _____
22. _____     47. _____
23. _____     48. _____
24. _____     49. _____
25. _____     50. _____

## Practical Examination Answer Sheet

1. _____
2. _____
3. _____
4. _____
5. _____
6. _____
7. _____
8. _____
9. _____
10. _____
11. _____
12. _____
13. _____
14. _____
15. _____
16. _____
17. _____
18. _____
19. _____
20. _____
21. _____
22. _____
23. _____
24. _____
25. _____

26. _____
27. _____
28. _____
29. _____
30. _____
31. _____
32. _____
33. _____
34. _____
35. _____
36. _____
37. _____
38. _____
39. _____
40. _____
41. _____
42. _____
43. _____
44. _____
45. _____
46. _____
47. _____
48. _____
49. _____
50. _____

# Credits

## Text and Line Art

### Laboratory 1
Laboratory adapted from Kathy Liu, "Eye to Eye with Garden Snails," Pacifica, CA.

### Laboratory 3
Figure 3.10: Originally in P. Anderson and S. Mader, *General Biology*, Copyright © 1973 Kendall/Hunt Publishing Company, Dubuque, Iowa.

### Laboratory 11
Figure 11.5, 11.6: From Peter H. Raven and George B. Johnson, *Biology*, 5th edition. Copyright © 1999 The McGraw-Hill Companies. All Rights Reserved.

### Laboratory 15
Figure 15.10: b. Modified from T. T. Storer and R. L. Usinger, *General Zoology*. Used with permission of California Academy of Sciences, Golden Gate Park, San Francisco, CA. 15.13: After Carolina Biological Supply.

### Laboratory 19
Figure 19.5: From David Shier, et al., *Hole's Human Anatomy & Physiology*, 10th edition. Copyright © 2004 The McGraw-Hill Companies. All Rights Reserved.

### Laboratory 20
Figure 20.14: Adapted from A. Glenn Richards, *The Complementarity of Structure and Function: A Laboratory Block*, 1969.

### Laboratory 21
Figure 21.3: Adapted from *American Biology Teacher*, April/May 1983:221. Drawn by Kristine A. Kohn.

## Photos

### Laboratory 1
Figure 1.1: © James Robinson/Animals Animals/Earth Scenes.

### Laboratory 2
Figure 2.3a: © Michael Ross/Photo Researchers, Inc.; 2.3b: © CNRI/SPL/ Photo Researchers, Inc.; 2.3c: © Steve Gschmeissner/SPL/Photo Researchers, Inc.; 2.4, 2.5: Courtesy of Leica; 2.9: © T.E. Adams/Visuals Unlimited.

### Laboratory 3
Figure 3.1: © Ed Reschke/Peter Arnold, Inc.; 3.2: © Ralph A. Slepecky/Visuals Unlimited; 3.5a,b: Courtesy Ray F. Evert, University of Wisconsin; 3.9a–c: © David M. Phillips/ Visuals Unlimited; 3.10a: © Dwight Kuhn; 3.10b: © Alfred Owczarzak/Biological Photo Service.

### Laboratory 7
Figure 7.2a: © Science Source/Photo Researchers, Inc.; 7.3(Prophase right & left), 7.3(all), 7.4(all): © Ed Reschke; 7.5(all): Photo Courtesy Dr. Andrew S. Bajer, University of Oregon; 7.6a(top, bottom): Copyright by R.G. Kessel and C.Y. Shih, *Scanning Electron Microscopy in Biology: A Students' Atlas on Biological Organization*, Springer-Verlag, 1974; 7.6b: © B.A. Palevitz and E.H. Newcomb/ BPS/Tom Stack & Associates.

### Laboratory 8
Figure 8.3a: © Superstock; 8.3b: © Michael Grecco/Stock Boston; 8.3c,d: © The McGraw Hill Companies, Inc./John Thomeing, photographer; 8.3e,f,h: © The McGraw-Hill Companies, Inc./Bob Coyle, photographer; 8.3g: Courtesy Mary Drapeau; 8.4: © CABISCO/Phototake.

### Laboratory 10
Figure 10.2c: © CNRI/SPL/Photo Researchers; 10.10a,b: © Bill Longcore/ Photo Researchers, Inc.

### Laboratory 11
Figure 11.4: © 2001 Time Inc., reprinted by permission.

### Laboratory 12
Figure 12.1: © Ralph A. Slepecky/Visuals Unlimited; 12.2a: © Alfred Pasieka/SPL/ Photo Researchers, Inc.; 12.2b: © David Scharf/SPL/Photo Researchers, Inc.; 12.2c: © Science Source/Photo Researchers, Inc.; 12.3a: © M.I. Walker/Photo Researchers, Inc.; 12.3b: © R. Knauft/ Science Scource/Photo Researchers, Inc.; 12.3c: © David Hall, University of London, Kings College/Photo Researchers, Inc.; 12.4b: © M.I. Walker/Photo Researchers, Inc.; 12.5c: © Carolina Biological Supply/ Phototake; 12.6(inset): © Dr. Anne Smith/ SPL/Photo Researchers, Inc.; 12.6(background): © John Cunningham/Visuals Unlimited; 12.7c: © Ed Reschke/Peter Arnold, Inc.; 12.9(right): © V. Duran/Visuals Unlimited; 12.9(left): © Cabisco/Visuals Unlimited; 12.10a: © Biophoto Assoc./Photo Researchers, Inc.; 12.10b: © Bill Keogh/ Visuals Unlimited; 12.10c: © Gary R. Robinson/Visuals Unlimited; 12.10d: © Michael Vivard/Peter Arnold; 12.11a: © Gary T. Cole/Biological Photo Service; 12.12(bread): © Precision Graphics/Kim Brucker, photographer; 12.12(mold): © James W. Richardson/ Visuals Unlimited; 12.13a,b: © Carolina Biological Supply/Phototake; 12.15a: © Everett S. Beneke/Visuals Unlimited; 12.15b: © John Hadfield/SPL/Photo Researchers, Inc.; 12.15c: © P. Marazzi/ SPL/Photo Researchers, Inc.

### Laboratory 13
Figure 13.1: © Stephen P. Lynch; 13.3a(Moss): © John Shaw/Tom Stack & Associates; 13.3a(inset): © Ed Reschke; 13.3b(Fern): © William E. Ferguson; 13.3b(inset): Courtesy Graham Kent; 13.3c(Conifer): © Kent Dannen/Photo Researchers, Inc.; 13.3c(inset): Courtesy Graham Kent; 13.3d(Cherry blossom): © E. Webber/Visuals Unlimited; 13.3d(inset): © Richard Shiell/Animals Animals/Earth Scenes; 13.7(Sword fern, Nephrolepis hirsatula): © McGraw-Hill Companies, Inc./Carlyn Iverson, photographer; 13.8: © Phototake; 13.10a: © Grant Heilman Photography, Inc.; 13.10b: © Walt Anderson/Visuals Unlimited; 13.10c: © James Mauseth; 13.12a: © J.R. Waaland/Biological Photo Service; 13.12b: © Ed Reschke; 13.13: © Carolina Biological Supply/Phototake.

### Laboratory 14
Figure 14.3b: © J.R. Waaland/Biological Photo Service; 14.5a: Courtesy Ray F. Evert, University of Wisconsin, Madison; 14.5b: © CABISCO/Phototake; 14.6: Courtesy John Selier Virginia Tech Forestry Department; 14.7: © Carolina Biological Supply/Phototake; 14.9a: © CABISCO/Phototake; 14.10: Courtesy Wilfred A. Cote, SUNY College of Environmental Forestry.

### Laboratory 15
Figure 15.1: © A & J Visage/Peter Arnold, Inc.; 15.3(Snail): © Bob Coyle/Red Diamond Stock Photo; 15.3(Nudibranch): © Kenneth W. Fink/Bruce Coleman; 15.3(Octopus): © Alex Kerstitch/Bruce Coleman; 15.3(Nautilus): © Douglas Faulkner/Photo Researchers; 15.3(Scallop): Courtesy of Larry S. Roberts; 15.3(Mussels): © Rick Harbo; 15.5b, 15.6b: © Ken Taylor Wildlife Images; 15.7a: © Kim Taylor/Bruce Coleman; 15.7b: © James Carmichael/ Nature Photographics; 15.7c: © Natural History Photographic Agency; 15.7d: © Kjell

Sandved/Butterfly Alphabet; **15.8b:** © Ken Taylor Wildlife Images; **15.9b(Scale):** © Science VU/Visuals Unlimited; **15.9b(Beetle):** © Kjell Sandved/Bruce Coleman; **15.9b(Lacewing):** © Glenn Oliver/Visuals Unlimited; **15.9b(Housefly):** © L. West/Bruce Coleman; **15.9b(Walking stick):** © Art Wolfe; **15.9b(Honeybee):** © John Shaw/Tom Stack & Associates; **15.9b(Flea):** © Edward S. Ross; **15.9b(Dragonfly):** © John Gerlach/Visuals Unlimited; **15.9b(Moth):** © Cleveland P. Hickman; **15.9b(Grasshopper):** © Gary Meszaros/Visuals Unlimited; **15.9b(Leaf hopper):** © Farley Bridges; **15.14a:** © Hal Beral/Visuals Unlimited; **15.14b:** © Patrice/Visuals Unlimited; **15.14c:** © Rod Planck/Tom Stack & Associates; **15.14d:** © Suzanne and Joseph Collins/Photo Researchers, Inc.; **15.14e:** © Robert and Linda Mitchell

Photography; **15.14f:** © Craig Lorenz/Photo Researchers, Inc.; **15.15:** © Rod Planck/Photo Researchers, Inc.; **15.16b:** © Carolina Biological Supply/Phototake; **15.17, 15.18, 15.19:** © Ken Taylor Wildlife Images.

## Laboratory 16

**Figures 16.3b, 16.5:** © Ken Taylor Wildlife Images.

## Laboratory 17

**Figures 17.2b, 17.3b:** Courtesy of Yvonne Szymanski, Niagara County Community College; **17.10:** Courtesy of Omron Healthcare.

## Laboratory 18

**Figure 18.1b:** © Ed Reschke/Peter Arnold, Inc.

## Laboratory 19

**Figures 19.4, 19.7:** © The McGraw Hill Companies, Inc./Carlyn Iverson, photographer; **19.9, 19.10:** © Ed Reschke; **19.11:** © Ken Taylor Wildlife Images.

## Laboratory 20

**Figures 20.2a–c:** Courtesy of Dr. Timothy Cannon, Ph.D; **20.5:** © Manfred Kage/Peter Arnold, Inc.; **20.6a:** Copyright © 2005 The Regents of the University of California. All Rights Reserved. Used by permission; **20.6b:** © PH Gerbier/SPL/Photo Researchers, Inc.; **20.7a(Housefly):** © L. West/Bruce Coleman; **20.7b(Nautilus):** © Douglas Faulkner/Photo Researchers, Inc.

## Laboratory 21

**Figure 21.3:** © NOAA/Visuals Unlimited.

# Index

Page numbers followed by an *f* or *t* represent figures or tables, respectively.

Interstitial cells, 250
Intestines
    of bivalve, 187f
    of clam, 186
    fat digestion in, 236–237
    of fetal pig, 215, 216f, 217
    of frog, 201, 202f, 203f, 204f
    of grasshopper, 195f
    role of, 215
    of squid, 189f
Inversion, 18
Invertebrates
    eyes of, 263
    vs. vertebrates, 205
Iris, 264f, 265t
Isopods, 2
Isotonic solution, 33

## J

Jacobs syndrome, 107f, 108
Jugular veins, 224f, 225

## K

Karyotypes, 106
Kidneys
    anatomy of, 243f
    of bivalve, 187f
    blood supply to, 222f
    of frog, 203f, 204f
    in urinary system, 242f, 243
Klinefelter syndrome, 107f, 108
Knee-jerk reflex, 262f, 263

## L

Labial palps, 186, 187f
Laboratory safety procedures, vii
Labrum, of grasshopper, 194f
Larynx
    of fetal pig, 214, 216f
    of frog, 202f
Leaf anatomy, 179f
Length
    of fingers, as inherited trait, 82f, 83t
    measurement of, 10–11, 10t
Lens, 265t
Ligament, hinge, 185f, 186
Light, in photosynthesis, 56–59
Light microscopes, 12
Lipase, 46
Lipids, pancreatic lipase digestion of,
        236–237
Liquid, diffusion through, 29
Liter, 10
Liver
    blood supply to, 222f
    digestive role of, 237f
    of fetal pig, 216f
    of frog, 202f, 204f

Locomotion
    by crayfish, 191f, 192
    by grasshopper, 193
    by mollusks, 184
    by squid, 188, 189f
Longitudinal section, 9
Luna moth, 193f
Lungs
    blood supply to, 222f
    of fetal pig, 211f, 216f
    of frog, 201, 202f, 203f, 204f
Lysosome, 25t

## M

Malphigian tubules, 195, 195f
Mammals
    brain of, 260
    defining characteristics of, 208
    habitats of, 198
    thoracic cavity in, 214
Mammary glands, 208
Mandible, of grasshopper, 194f
Mantle
    of bivalve, 187f
    of clam, 186
    of mollusks, 184
    of squid, 189f
Mass, measurement of, 10–11
Maxilla, of grasshopper, 194f
Maxillary palp, 194f
Maxillary teeth, of frog, 200, 200f
Measurement
    of meniscus, 11
    metric system, 10–11
Medulla oblongata, 258, 259f, 260f
Megasporangia, 160
Meiosis, 63
    crossing-over in, 72
    homologues in, 71
    vs. mitosis, 77f
    nondisjunction in, 107
        during phase I, 109–110
        during phase II, 110
    phases of, 74f–75f
    in plants
        in alternation of generations, 155f
        ferns, 158f
        flowering, 164f
        mosses, 156f
        pine tree, 161f
Meniscus, 11
Meristem, 170
Mesenteries, 215
Mesonephric ducts, 202, 203f, 204f
Mesothorax, 193
Mesozoic era, 122t, 123, 124f, 125f
Messenger RNA (mRNA), 96
    in transcription process, 97f
    in translation process, 98
Metabolism, 39

Metamorphosis
    definition of, 196
    in insects, 196, 197f
Metaphase, 67f, 74f, 75f
Metathorax, 193f
Meter, 10t
Methylene blue, 20
Metric system, 10–11
Micrographs, comparative, 12f
Micrometer, 10t
Microscopy
    binocular dissecting microscope, 14–15
    compound light microscope, 12
        focusing, 17–19
        parts of, 16–17
        vs. transmission electron
            microscope, 13t
    electron microscopes, 13
    inversion in, 18
    light microscopes, 12
    total magnification in, 19
    wet mount preparation for, 19
Microsporangia, 160
Midbrain, 258, 259f, 260
Middle ear, 268f
Milliliters, 10
Millimeter, 10t
Mitochondrion, 25t
Mitosis, 63
    vs. meiosis, 77f
    phases of
        in animal cells, 66f–67f
        in plant cells, 69–70, 69f
    structures associated with, 64t
Mitotic spindle, 64t, 65
Mold, on bread, 148f, 149–150
Molecular genetics, 91
Molecules, ATP, 47
Mollusks
    classes of, 184
    locomotion in, 184
Monocot plants, 169, 171f
Monohybrid cross, 80f
Monosaccharides, 48
Morel, 148f
Mosses, 154f, 156–157, 156f
Moth, Luna, 193f
Motor neuron, 257f
Mouth
    of crayfish, 191f
    of fetal pig, 210
    of grasshopper, 194f, 195f
    of squid, 189f
Multicellularity, in animal
        evolution, 182f
Muscle(s)
    adductor
        of bivalves, 187f
        of clams, 186
    ciliary, 264f, 265t
    effector, 261f

Muscular dystrophy, 116t
Mussel, 184, 184f
Mycelium, 148
Myelin sheath, 257f

## N

Nares, of frog, 199, 199f
Nasopharynx, 210, 210f
Near point, 266
Nephrons, 243f
Nerves, spinal, 261
Nervous system, 257
Net photosynthesis, 56
Neurofibromatosis, 114, 116t
Nictitating membrane, 199
Nipples, 208
Nondisjunction, 107
    simulating
        during meiosis I, 109–110
        during meiosis II, 110
Notochord, 198
Nuclear envelope, 25
Nuclei, 24
Nucleoid, 24
Nucleoli, 63
Nucleolus, 25t, 64t
Nucleoplasm, 25, 63
Nucleus, 25t, 64t

## O

Occipital lobe, 258, 259f
Octopus, 184
Olfactory bulb, 259f
One-trait crosses, 80–81
Oocyte, 249f, 250, 251f
Oogenesis, 109
Optic chiasma, 259f
Optic nerve, 264f, 265t
Oral thrush, 113
Organelles, 24, 25
Organization level, in animal
        classification, 183t
Oscillatoria, 142f
Osmosis, 32–35, 32f
Ossicles, 268f
Otic vesicles, 267
Outer ear, 268f
Ovarian tissue, 251f
Ovaries, 241
    of fetal pig, 247, 248f
    of frog, 201, 202f, 204f
Oviducts
    of fetal pig, 247, 248f
    of frog, 204f
Ovipositor, 194, 194f
Ovule, 160, 165

## P

Palate, 210, 210f
Paleozoic era, 122t, 123, 124f, 125f

Pancreas
    in fat digestion, 237f
    of fetal pig, 217
    of frog, 201
    role of, 215
Pancreatic lipase, fat digestion by,
        236–237
Paper chromatography, 54f
Paramecium, 145f
Parfocal, 18
Parietal lobe, 258, 259f
Pathogenic bacteria, 137–138
Pedigrees, of genetic disorders, 111–112
Penis
    of fetal pig, 209, 244, 245f
    location of, pig vs. human, 246f
Pepsin, protein digestion by, 234–235
Peptidoglycan, 141
Pericardium
    of bivalve, 187f
    of clam, 186
    of frog, 201
Peritoneal cavity, 243
Peritoneum, 215, 243
Peroxisome, 25t
Petal, 165
pH
    in enzymatic activity, 44–45
    in enzyme denaturation, 40
    in fat digestion test, 236
    in stomach, 234
pH scale, 36
Pharyngeal pouches, 198
Pharynx, 210, 211, 211f
Phenol red, 236
Phenotype, 79, 111
Phenylketonuria (PKU), 114, 116t
Phloem, 173, 176
Photomicrograph, 12
    of cyanobacterium, 142f
    of epithelial cell, 23f
    of ovarian tissue, 251f
    of seminiferous tubule, 250f
    of spinal cord, 262
Photoreceptors, 264
Photosynthesis
    action spectrum for, 58f
    in carbon cycle, 61
    carbon dioxide uptake in, 60–61
    equation for, 53
    green light in, 58–59
    gross, 57
    net, 56
    overview of, 53f
    plant pigments in, 54–55
    rate of, 57
    white light in, 56–57
Pig. See Fetal pig
Pillbugs, 1–2, 1f
Pine cones, 162–163, 162f
Pine tree life cycle, 161f
Pineal gland, 260f

Pituitary gland, 260f
Placenta, 208
Plants
    adaptation by, 153
    alternation of generations in, 155
    cell structure of, 25t
    comparison of, 167
    evolution of, 154f
    fossil record of, 125f
    meiosis in
        alternation of generations, 155f
        ferns, 158f
        flowering, 164f
        mosses, 156f
        pine tree, 161f
    mitosis in, 69–70, 69f
    pigments of, in photosynthesis,
        54–55
    seedless
        ferns, 158–159, 158f
        mosses, 156–157, 156f
    with seeds
        angiosperms. See Flowering plants
        gymnosperms, 161f, 162–163
        leaf anatomy of, 179f
        leaf organization of, 178–179
        monocots vs. eudicots, 171f
        organs of, 170
        root organization of, 172–173
        sexual reproduction in, 160f
        stem organization of, 175–177
        twig structure, 176f
        xylem transport in, 174
    tree rings, 177f
Plasma membrane
    diffusion across, 30–31
    of elodea cell, 35f
    function of, 23, 25t
    of prokaryotic cell, 24f
Plasmids, 138
Plasmodial slime molds, 147
Plasmolysis, 34
Pollen grain, 160
Pollen tube, 165
Pollination, 160, 165
Pollutants, 273
    effects of
        on Gammarus, 275–276
        on hay infusion cultures, 274–275
        on seed germination, 275
Polymerase
    DNA, 101
    RNA, 97
Polymerase chain reaction (PCR), 101
    in DNA fingerprinting, 102f
Polypeptides
    amino acid sequence in, 98
    growth of, 99f
Polyplacophora class, mollusks in, 184
poly-X syndrome, 107f, 108
Pond water, protists in, 146
Pons, 258, 259f, 260f

Spermatogenesis, 109
Spinal cord
    anatomy of, 261
    gray and white matter in, 262
    in reflex arc, 261*f*
    in vertebrates, 198
Spinal nerves, 261
Spinal reflexes, 261*f*, 262–263
Spindle, 64*t*, 65
Spiracles, 194, 194*f*
*Spirilla*, 140
*Spirogyra*, 143*f*
Spleen, 215
Sporangium, 147, 160
Spores, 148, 155
Sporophyte, 155
Sporozoans, 145
Squid
    anatomy of, 188, 189*f*
    eye of, 263
    locomotion by, 188, 189*f*
Stamen, 165
*Staphylococcus*, 140
Starch, salivary amylase digestion of,
    238–239
Stems, of flowering plants, 175–177
Stereomicroscope, 14–15
Stomach. *See also* Digestive system
    of bivalve, 187*f*
    of fetal pig, 215
    of frog, 201, 202*f*, 203*f*
    of grasshopper, 195*f*
    pH in, 234
    protein digestion in, 234*f*
    role of, 215
    of squid, 189*f*
Stomata, 169, 178*f*
*Streptococci*, 140*f*
Substrates, 39
Sucrose, 48
Swimmerets, 191*f*, 192
Symmetry, in animal evolution,
    182*f*, 183*t*
Syndrome, 105
Systemic circuit, 222*f*, 223, 225
Systolic blood pressure, 229

## T

*T. pallidum*, 140*f*
Taste buds, 271
Tay-Sachs disease, 116*t*
Teeth
    of fetal pig, 210, 210*f*
    maxillary, 200, 200*f*
    vomerine, 200, 200*f*
Telophase, 67*f*, 74*f*, 75*f*
Telson, 191*f*
Temperature
    effect of, on enzyme activity, 42
    sensory receptors for, 271
Temporal lobe, 258, 259*f*

Tendon, central, 212
Testes, 241
    of fetal pig, 244, 245*f*
    of frog, 202, 203*f*
    function of, 244*t*
    of human, 246*f*
Tetrapods, 208
Thalamus, 258, 259*f*, 260*f*
Thermal pollution, 275, 276
Thistle tube, 32*f*
Thoracic cavity, 212, 214, 216*f*
    veins of, 226*f*
Thorax, of grasshopper, 193
Thrush, oral, 113
Thumb hyperextension, 83*t*
Thylakoids, 24
Thymus gland, 214
Thyroid gland, 211*f*, 214, 216*f*
Tissues, true, in animal
    evolution, 182*f*
Tobacco seedling color, 80–81
Tongue, 210*f*
Tonicity, 33–35
Touch receptors, 270*f*
Tracheae
    of fetal pig, 211*f*
    of grasshopper, 195
Transcription process, 96, 97
Transfer RNA (tRNA), 96*f*, 98, 99*f*
Translation process, 96, 98–99
Transmission electron micrograph, 12*f*
Transmission electron microscope, 13
Transpiration, 169
Tree rings, 177*f*
Tricuspid valve, 221
Triplet code, 98, 133
True tissues, in animal evolution, 182*f*
*Trypanosoma*, 145*f*
Tube-within-a-tube body plan, 183*t*
Turgor pressure, 34
Turner syndrome, 107*f*, 108
Turtle, Pearl River redbelly, 198*f*
Twig structure, 176*f*
Tympanic membrane, 268*f*
Tympanum, 194, 194*f*

## U

Umbilical arteries, 216*f*
Umbilical cord
    of fetal pig, 216*f*
    of mammals, 208
Umbilical vein, 212, 216*f*
Umbo (beak), 185*f*, 186, 187*f*
Ureters, 242*f*, 243, 246*f*
Urethra, 242*f*, 243
Urinary bladder, 242*f*, 243, 246*f*
Urinary system, 241
    of fetal pig, 242*f*, 243
*Urocyon cinereoargenteus*, 198*f*
Urogenital opening, 209*f*
Urogenital papilla, 209*f*

Urogenital system, 241
    of frog, 203*f*
    of human, 246*f*
Uropods, 191*f*, 192
Uterine horns, 247, 248*f*
Uterus
    of fetal pig, 247, 248*f*
    of frog, 204*f*
Uvula, 210

## V

Vacuoles, 145
Vagina, 247, 248*f*, 249*f*
Valves, of clams, 185
Vasa deferentia, 244, 245*f*, 246*f*
Vasa efferentia, 202, 203*f*
Vascular cambium, 176
Veins
    of eye, 264*f*
    of fetal pig, 224*f*, 226*f*, 228*f*
    hepatic, 255
    iliac, 227*f*
    renal, 227*f*, 243*f*
Ventricles, 258, 259*f*
Vertebrates, 198
    brain comparison of, 260
    embryo comparison of, 132
    forelimb comparison of, 127, 128*f*
    *vs.* invertebrates, 205
Vesicle, 25*t*
Visceral mass
    of clam, 186
    of mollusks, 184
Vision, 264, 265. *See also* Eye
Volume measurement, 10–11
Volumeter, 51*f*
*Volvox*, 144
Vomerine teeth, 200, 200*f*

## W

Walking stick, 193*f*
Water eutrophication, 273
Wet mount preparation, 19
White light, in photosynthesis, 56–57
White matter, 261*f*, 262
Widow's peak, 82*f*, 83*t*
Wings, of grasshopper, 193, 194*f*
Wood, 176

## X

Xanthophyllis, 54
X-linked genes, 86–89
    in color blindness, 114
Xylem, 169, 173, 176
Xylem transport, 174

## Y

Yeast fermentation, 47